APPROPRIATE VISIONS

APPROPRIATE VISIONS

Technology the Environment and the Individual

Edited by
Richard C. Dorf
and
Yvonne L. Hunter

University of California, Davis

BOYD & FRASER PUBLISHING COMPANY
SAN FRANCISCO

Richard C. Dorf and Yvonne L. Hunter, editors
Appropriate Visions: Technology, the Environment, and the Individual

Library of Congress Cataloging in Publication Data:
Main entry under title:
Appropriate visions.
Bibliography: p.
1. Technology—Addresses, essays, lectures. 2. Technology—Social aspects—Addresses, essays, lectures. 3. Human ecology—Addresses, essays, lectures.
I. Dorf, Richard C. II. Hunter, Yvonne L.
T185.A68 301.24'3 78-9045
ISBN 0-87835-072-1
ISBN 0-87835-069-1 pbk.

1 2 3 4 5 · 2 1 0 9 8

CONTENTS

PREFACE

Technology is science plus purpose. However, the consequences of our technologies are often unforeseen and undesirable. The thoughts expressed in this book reflect concerns with the impact of technologies on our relationship with the environment, to other persons, and, most importantly, to ourselves.

In late 1975, the Yolo County Chapter of the American Civil Liberties Union and the Unitarian Universalist Association of Churches joined with the University of California, Davis, to plan and sponsor a series of programs which would address the problems of growing technology, the need for energy conservation, and the maintenance of human values. The two events that resulted from this planning were an address by Barry Commoner on October 23, 1976, and a two-day conference featuring E. F. Schumacher, titled "Small Is Beautiful." It was held on February 17 and 18, 1977.

This book is essentially a written record of those events and the visions of Commoner, Schumacher, and others. Regrettably, some of the excitement, spontaneity, euphoria, and intellectual creativity that surrounded the programs may be lost. For example, the Schumacher conference included an exhibit of practical intermediate technology projects from around California and an energy conservation tour of the city of Davis. We trust that the photos included in the book will provide some feeling of the vibrant atmosphere that encompassed the programs.

The purpose of this book is to assist, in a small way, in restoring that splendid spirit of self-reliance we seem to have let fade away and to work towards a world where the individual and human values are in tune with the visions that have come to us from Barry Commoner and E. F. Schumacher.

Commoner and Schumacher point us, anew, to the development of technology that will be appropriate to the task, conform to local requirements, and permit socially acceptable forms of growth. Without human and social developments, they correctly assert, technology can achieve little. Schumacher enjoyed quoting Thomas Aquinas, who said, "The slenderest knowledge that may be obtained of the highest things is more desirable than the most certain knowledge of lesser things." Both Commoner and Schumacher, as well as other panelists, continually reflected on our misplaced, rationalistic faith in science and technology. They help us, in part, to repudiate scientism, to reclaim the value of wisdom, and, hopefully, to adopt a more holistic view of the world that includes the individual and technology. Commoner quite clearly points out that inappropriate or immature technology is never acceptable. Schumacher says that the only answer to the perplexing question, "What can I actually do?" is to "work to put our own house in order."

Appropriate Visions is divided into eight sections. In the first, Commoner reflects on freedom, the ecological imperative, and the use of energy and technology. The second begins with an introduction to E. F. Schumacher and his thoughts on intermediate technology and the individual. It is followed successively by sections on agriculture, the ethical basis of intermediate technology, and Third World development. Reflections on changing the scale of technology are followed by a consideration of energy and intermediate technology. Finally, a summary and conclusion section is provided.

The reader will note by examining the list of contributors that Commoner and Schumacher are joined by a large number of distinguished panelists from a range of disciplines, ideologies, and backgrounds. Their insightful contributions enriched the discussions.

A number of projects now under way at the University of California, Davis, received impetus from the Commoner and Schumacher programs. The University's new Institute of Appropriate Technology will provide leadership in the development of research, teaching, and information dissemination in the field of appropriate technology. The University of California, Davis, is the campus responsible for the new University-wide Energy Extension Service, which will develop energy conservation information for a wide range of audiences. The student-initiated Agricultural Alternatives Development Program is receiving full support and cooperation from students, faculty, and deans.

Programs of the magnitude of the Schumacher and Commoner events cannot be successfully accomplished by two people alone. There are numerous individuals to whom we owe our thanks. The members of the Program Steering Committee from the ACLU and the Unitarian church offered valuable suggestions throughout the planning stages. They are: Mitzi Ayala, Evelyn Rominger, Art Small, Fran DuBois, Bob Lieber, Larry Hoover, Charles Slap, Bruce Taylor, and Pauline Asher. Sandy Goff and Peter Gillingham of Intermediate Technology in Menlo Park were a delight to work with during the Schumacher Conference. Ned Engle and Henry Esbenshade supplied good humor and recommendations throughout and were responsible for putting together the Intermediate Technology Fair. The expert professional assistance of the staff of University Extension, especially Pat Erigero, Gayle BonDurant, Philip Cecchetini, and Carol Wageman, helped us to transform our ideas into reality. In addition, we are indebted to Phylis Needle and Edi Weaver who had the monumental task of transcribing the manuscripts from the tape recordings of the programs.

Finally, we would like to dedicate this book to the work and wisdom of Fritz Schumacher, who died suddenly in September, 1977. The wisdom and power of his message has enriched us all; its strength will carry on.

RICHARD C. DORF
YVONNE L. HUNTER

Davis, California
November, 1977

CONTRIBUTORS

TOM BENDER, Editor, *RAIN* magazine, Portland, Oregon

VASHEK CERVINKA, Senior Analyst, California Department of Food and Agriculture

JOHN COLEMAN, S.J., Assistant Professor of Religion and Society, Jesuit School of Theology and the Graduate Theological Union, Berkeley, California

BARRY COMMONER, Director, Center for Biology of Natural Systems, Washington University, St. Louis, Missouri

PAUL CRAIG, Director, UC Council on Energy and Resources; Professor of Applied Science, University of California, Davis

RICHARD C. DORF, Dean, Division of Extended Learning, University of California, Davis

FRANCIS DUBOIS, Chairman, Demeter Corporation, Woodland, California

RANDALL FIELDS, President and Chairman, Fields Consulting Group, Menlo Park, California

ISAO FUJIMOTO, Lecturer, Department of Applied Behavioral Sciences, University of California, Davis

ROGER GARRETT, Professor, Department of Agricultural Engineering, University of California, Davis

PETER GILLINGHAM, Director, Intermediate Technology, Menlo Park, California

GARY GOODPASTER, Chief Assistant Public Defender, State of California; Professor of Law, University of California, Davis

JONATHAN HAMMOND, Director, Living Systems, Winters, California

GARRETT HARDIN, Professor of Human Ecology, Department of Biological Sciences, University of California, Santa Barbara

YVONNE L. HUNTER, Assistant to the Administrator, Kellogg Program, University of California, Davis

DESMOND JOLLY, Cooperative Extension Consumer Specialist and Lecturer, Agricultural Economics, University of California, Davis

JOHN KEMPER, Dean, College of Engineering; Professor of Mechanical Engineering, University of California, Davis

HERMAN KOENIG, Director, Center for Environmental Quality; Professor of Electrical Engineering and Systems Science, Michigan State University, East Lansing, Michigan

E. PHILLIP LEVEEN, California Agricultural Experiment Station, Giannini Foundation of Agricultural Economics, University of California, Berkeley

LOUIS B. LUNDBORG, Chairman of the Board (retired), Bank of America, San Francisco, California

ALEX MCCALLA, Professor, Department of Agricultural Economics, University of California, Davis

GLORIA MACGREGOR, Director of Planning, City of Davis, California

P. K. MEHTA, Professor, Department of Civil Engineering, University of California, Berkeley

DONALD R. NEILSEN, Associate Dean, College of Agriculture and Environmental Sciences; Chairperson, Department of Land, Air, and Water Resources, University of California, Davis

MICHAEL PERELMAN, Department of Agricultural Economics, Chico State University, Chico, California

RICHARD ROMINGER, Director, California Department of Food and Agriculture; Yolo County farmer

E. F. SCHUMACHER, Founder, Intermediate Technology Development Group, London

STEVEN SLABY, Director, Technology and Society Seminars, College of Engineering and Applied Science, Princeton University, Princeton, New Jersey

SIM VAN DER RYN, State Architect; Director, California Office of Appropriate Technology, Sacramento, California

LYMAN P. VAN SLYKE, Assistant Professor, Department of History, Stanford University, Stanford, California

EMILIO VARANINI, Commissioner, California State Energy Resources Conservation and Development Commission, Sacramento, California

KENNETH E. F. WATT, Professor of Zoology, Department of Zoology, University of California, Davis

DAN WHITNEY, Senior Nuclear Engineer, Rancho Seco Nuclear Generating Station, Sacramento Municipal Utility District, Sacramento, California

Beginning left, clockwise: Barry Commoner; Steven Slaby; E. F. Schumacher; Agriculture Panel: (left to right) Peter Gillingham, Michael Perelman, Richard Rominger, Roger Garrett, Vashek Cervinka, Phillip LeVeen.

Beginning left, clockwise:
E. F. Schumacher; Louis Lundborg and Herman Koenig; Panel: (left to right) E. F. Schumacher, John Kemper, Yvonne Hunter, Richard Dorf; Garrett Hardin.

I

FREEDOM AND THE ECOLOGICAL IMPERATIVE

Barry Commoner

INTRODUCTION: *Richard C. Dorf*
MODERATOR: *Francis DuBois*
COMMENTATORS: *Gloria MacGregor*
Randall Fields

Introduction to Barry Commoner

Richard C. Dorf

There is an increasing sense among the people of the United States that our freedom is eroding, that it has become more limited as the government's power, often without the consent of the governed, has increased. Many feel that our liberty has become more tenuous as the government's power, isolated from the people's will by an elaborate bureaucracy, has grown more absolute. Our freedom, which we won two hundred years ago, is not threatened by a foreign monarchy, but by the very government which we ourselves have created and unleashed—one that we support with our own earnings, one that we have empowered with laws enacted by our own representatives.

Dr. Barry Commoner, one of the most influential and insightful environmentalists of 1970's, has exposed the very positive and negative aspects of the environmental movement in the following paper. Dr. Commoner was educated at Columbia University where he received his bachelor's degree, and at Harvard University where he received his master's and Ph.D. in biology in 1938 and 1941, respectively. He also received an honorary doctorate at the University of California, among others. During World War II, Dr. Commoner served as a Naval Air Force Observer. He joined Washington University at St. Louis as Associate Professor of Plant Physiology in 1947 and has been with that University ever since. He became the Director of the Center for the Biology of Natural Systems in 1965 and continues in that position. Dr. Commoner has served for a number of years in the development and leadership of the Scientists' Institute for Public Information activities. Since

1954, he has also held a number of positions in the American Association for the Advancement of Science.

Dr. Commoner is an environmentalist with an insight into economics, a citizen who understands ecology, the production system, and economics. He is one who has attempted, as very few have, to examine the interrelationship of the ecological, economic, and political systems. In this instance, Dr. Commoner has looked at the number of pervasive new laws, among them the National Environmental Policy Act, the Clean Air Act, the Water Quality Improvement Act, and the Occupational Safety and Health Act, which have created a new system of regulation and growing bureaucratic machinery that set new limits on not only environmental degradation, but also on human freedoms. The right to enforce these laws and to write the regulations has resulted in new powers for the government and new losses of freedom for the individual. As new hazards and new thresholds of illness are identified, new regulations are written to avoid the dangers. Thus, as Commoner sees it, the major outcome of the national effort to improve the environment has been the creation of many new legislative constraints and a rapidly growing superstructure of regulations, reports, inspection, monitoring, hearings, civil and criminal proceedings, suits and countersuits, court orders, injunctions, and fines. No matter how much we value the improvement of the environment, realism requires that we recognize these outcomes of our environmental concerns for what, in truth, they are: stringent and pervasive limitations on freedom. The new bureaucracy has a new weapon. It is called "protection of the environment."

In our zeal to defend the environment we have added to the heavy and growing burdens that constrain our freedom. This tangled knot of issues involving environmental protection and individual freedom continues to frustrate the political process. As long as it remains knotted, belief in the value of the democracy that we created two hundred years ago is discouraged. A faulty design, in Dr. Commoner's view, has been imposed on the production system by the economic system, which itself is ultimately faulty. This faulty economic system, in his view, invests in operations that promise increased profits, rather than environmental compatibility and the efficient use of resources. If we know that we cannot survive without a balance between environmental quality, reasonable economic stability, and a production system that can deliver goods, food and services that we require, what then is the reality of the situation and the cause of the failure? How can we say the economic system has failed, and how can we propose that it improve? These are the questions that Dr. Commoner has raised in his books and discusses in the following paper.

Dr. Commoner is the author of three well-known and important books. His first book, entitled *Science and Survival,* appeared in 1966. His second book, *The Closing Circle,* was published in 1971, while his third, and most recent work, *The Poverty of Power,* was published in 1976. Dr. Commoner is one of the most widely read authors in the United States in the area of science and public policy. He is one of C. P. Snow's "men for all seasons" and is a charter member of the two cultures.

Commoner's most recent book, *The Poverty of Power,* subtitled "Energy and the Economic Crisis," is concerned with the elucidation of thermodynamics and its role in the economy of energy. In his earlier book, *The Closing Circle,* Commoner argued that, by breaking up the circle of renewable life systems, we have been plunging towards irreversible ecological catastrophe. In *The Poverty of Power* he extends his argument, stressing again the interdependence of every element within an ecological or economic system and then demonstrating how these systems are themselves interdependent and inescapably linked to other systems of business and government. In the last five years, matters have brought the environmentalists up against fundamental economic and political issues which include abatement of pollution by fossil fuels, which may mean the transformation of the fuel base of the economy of the United States. As a scientist, Dr. Commoner continues to believe that the revitalization of the United States is as much a technical problem as a political problem. Therefore, he continues to argue that technical, as well as social, revitalization is necessary.

Dr. Commoner analyzes the world, then, as well as the United States, in terms of three interlocking systems: the ecosystem, the production system, and the economic system. He quite adequately points out their complex interdependency and notes the malfunctions which result in broad ramifications. The primary ramification in the area of the ecosystem is the energy waste resulting from our affluent system of energy use. Until the solar industry matures, Dr.

Commoner sees our conventional fossil fuels, particularly petroleum and coal, as the answer to our energy needs. As he said in his earlier book, *The Closing Circle,*

> Anyone who presumes to explain a serious problem is expected to offer to solve it as well, but none of us . . . can possibly blueprint a specific plan for resolving the environmental crisis. To pretend otherwise is only to evade the real meaning of the environmental crisis: that the world is being carried to the brink of ecological disaster not by a singular fault which some clever scheme can correct but by the phalanx of powerful economic, political and social forces that constitute the march of history.

Dr. Commoner has seen, over the last few years, the new hope for the United States as the use of solar energy. In his words, "The chief practical purpose of thermodynamics is to learn how energy can best be harnessed to work-requiring tasks." As OPEC has reminded the world, fossil fuels are both exhaustible and subject to political manipulation. Also, in Commoner's view, nuclear power is no answer to potential energy shortage. Therefore, he states that the answer is the continual exploitation of oil sources in the near future and solar sources in the long run. That is the answer to his technical question. The answer to the political question is, quite simply, *socialism.* Socialism in his view is not Russian socialism or Cuban Marxism but a wiser, new American political system based on public control of the means of production. Dr. Commoner's view is that the American economy is doomed unless major changes are made, for profits are the only approach under the current economic system. Commoner, in *The Poverty of Power,* states:

> Now all this has culminated in the ignominious confession of those who hold the power: That the capitalist economic system which has loudly proclaimed itself the best means of assuring the rising standard of living for the people of the United States, can now survive, if at all, only by reducing that standard. The powerful have confessed to the poverty of their power.
>
> No one can escape the momentous consequences of this confession. No one can escape the duty to understand the origin of this historic default and to transform it from a threat to social progress into a signal for a new advance.

In the final analysis, the energy crisis is an economic crisis. According to Commoner, the question is not so much will we have enough fuel, but will be able to afford the fuel we have available? In his most recent book, *The Poverty of Power,* he boils it down to this: The second law of thermodynamics which, in formal terms, holds that the entropy of the universe is constantly increasing, tells us that it is not exactly energy that should be conserved. After all, energy is never consumed, it is merely transformed. The second law tells us that high temperature and, therefore, high quality energy such as that released from an oil burner, should not be used to heat a home. That task is better left to a low temperature energy source such as solar heating systems. In other words, don't use a sledgehammer to drive a tack; you'll waste energy and damage the surroundings as you do. But, warns Commoner, "These considerations suggest that unless some countervailing force intervenes, the postwar transformations in production technology will tend to generate a shortage of both capital and jobs. . . . What is now threatened is the economic system itself." The solution of the immediate problem, in his view, is oil. He proposes that the exploitation of domestic oil over the next twenty to thirty years will allow the United States to develop a valuable solar energy source and an important conservation system which will allow us to have an energy system for the future.

Dr. Commoner argues that the United States reserve of oil is larger by a factor of two or three than that currently estimated; therefore, instead of providing a ten-year or twenty-year source, it may well result in a fifty-year domestic source. This, then, gives America, with its coal resources, a chance to shift totally to solar energy over the next thirty years with marginal dependence on the OPEC sources.

Whether the price of energy is over-regulated or too cheap is difficult to know, for the arguments continue. If energy is too cheap, and we should increase our price, we have in the past played into the sheiks' hands. If energy is becoming too expensive, we are burdening our poor. If we are to continue the development of a nuclear power industry, we must have capital, and from whence is it to come? These questions are difficult to answer, but Barry Commoner does not wince in the face of them. In his 1971 best-seller, *The Closing Circle,* Dr. Commoner states that the industry must be restructured to conform to ecology's unbending laws; specifically, he recommended that polluting products such as detergents or synthetic textiles be replaced by "good old natural ones" such as soap or cotton and wool. Just how to accomplish such a major change in industrial direction is not easy to understand. However, as Dr. Commoner is one of the most adept of raisers of questions, he also points us to the fact that fully 85 percent of the work potential in the nation's oil, coal, and uranium is wasted, in the sense that it is not used as work. Again, we are reminded of the three interlocking problems of energy, environment, and the economy. Nearly all of today's problems stem from two enormously difficult economic factors: unemployment and inflation. It was heretofore thought that as unemployment increased, inflation would drop. The breaking of that relationship has resulted, in 1976 and 1977, in increased inflation and increased unemployment. Therefore, Commoner, among others, proposes that the key to the unemployment and inflation problems may be a new energy policy recognizing the new realities. About 96 percent of United States energy comes from petroleum, natural gas, coal, and uranium, which are all nonrenewable resources and, therefore, tied to inflationary costs. As these resources are exploited and exhausted, their costs will continue to rise, which explains the continued cost of energy.

As we recognize that life often incorporates a balance between improvements in the enviroment and penalties in the economic and political sectors, we recognize that our human freedoms may be traded for improved quality of the environment, at least in the short run. In Dr. Commoner's view, between the two trade-offs in our lives (that is, the ecosystem on the left hand and the economic system on our right), is the production system in the middle. It is this very production system that, in sum, we must focus upon. It is this system that inappropriately produces and utilizes synthetic products rather than natural products; for example, plastics rather than leather. Again, it is this very inappropriate economic or production system that yields the nuclear power generators rather than solar conversion systems. In his view, ecology is linked to economics by changes in the production system. As he states, the capital shortage has brought us to a point where the vaunted claim that the American capitalist system could raise the standard of living and make a good profit at the same time has now been given up. It is not clear who has given it up, but, as Dr. Commoner says, if we are going to recognize that it is the character of the production system which governs this relationship, in his view, between the ecosystem and the rest of our lives, then the most important question is: Who is in control of this production system? It is a question which challenges the fundamental precepts of the capitalist system, namely: Anyone who owns capital ought to be free to invest it in whatever way gives him the best profit. It raises the issue of whether, for the sake of the environment and a stable economy, we have to give the people control over production.

Here we have Dr. Commoner's most cogent analysis and his statement of his thesis, which is that socialism will occur and it will spring from disillusionment with corporate greed. Again, we are unclear whether Commoner is a conservationist dedicated to controlling waste or a consumerist dedicated to limiting prices and promoting high levels of consumption. Perhaps a system can be developed which would result in both, but one does not see that immediately before him. As one review of *The Poverty of Power* states:

> This more conventional approach can be criticized, of course, for having been tried in this country before and for not having lived up to expectations. But there is no news in that. Marxist dogma, too, has failed to live up to expectations in its many implementations. Indeed, socialist economic systems exhibit many of the same failures, in environmental terms, as capitalist systems. The reply to the criticism is that the failures and crises to which Commoner refers are not necessarily failures of the capitalist system *per se,* but are merely failures to operate the system competently.

Very few people in our country today have provided the analysis and insight that Dr. Commoner has into the fact that, as has often been stated, everything is linked to everything else. Why we have moved from the easy and soft technologies of solar and wind and wood and the natural textiles of cotton and wool to the synthetics, the nuclear power, the oil and coal, is not easy to answer. However, Dr. Commoner reminds us that we can again move back in a direction of the softer, and, perhaps, more ecologically sound technologies. Whether these moves would require a socialistic system in order to achieve these beneficial and more pleasant aspects of life, is unsure. Perhaps, with the right incentives within our current system and the appropriate national policies we could achieve the same end.

As Cameron and Craig point out in their review:

> *The Poverty of Power* is a deeply provocative book. It provides new perspectives and in doing so has enriched our debate. Yet, because of Commoner's overestimate of current petroleum supplies, he deemphasizes the energy crisis and characterizes the problems we have at present as being intrinsic to our economic system. His conclusion that we should move, perhaps sharply, in the direction of a socialist economy may be correct, but he has not carried his burden of proof on this point.

It is certainly not clear today, and it would be difficult to know whether Dr. Commoner's insight has shown us the way for the 1980's for a number of years. Nevertheless, here is a man who has touched our concerns for environmental protection and energy availability, on the one hand, and human rights and the quality of life on the other.

Freedom and the Ecological Imperative: Beyond the Poverty of Power

Barry Commoner

I was most intrigued by the topic that you decided to center this symposium of talks around, because very often people have ignored human beings in thinking about ecology. I think that it's an extremely important and a very salutary thing that you are thinking about, the issue of people, their values, their rights, and their aims in connection with the ecosystem, because the ecosystem seems to be something autonomous; in fact, it is. It was there before we were, and we seem to be intruders. There are ecologists, or shall we say, ecologically minded people, who very often think that man is the worst, most intrusive species in the entire ecosystem and that he has to be put in his place, his population controlled, and slapped down and told not to do too much to "disturb the whooping crane." Now, this is a very serious issue, because it has enormous political implications for the way in which people live.

Therefore, I want to talk about the apparent conflict between what most of us regard as freedom, and, so to speak, minding the ecological store—doing the right thing for the environment. I want to put it to you very simply—what we have done in the name of ecology in the United States in the last five years or so has put heavy constraints on our freedom. I say that flatly, and for the next two minutes I will sound like Ronald Reagan, because it is a fact that when you establish a new bureaucracy such as the Environmental Protection Agency it neither spins nor weaves nor produces anything; all it does is to keep people from doing things that they want to do, like polluting the environment. It makes measurements of pollution in the air, it tells the auto companies that they ought to put things on the tail end of the car, it fines people, hauls them into court—these are all constraints on our freedom, and I think that the issue of bureaucracy, that is, an elaborate system of controlling what people do, is one that cannot be ducked. The fact that control over what people want to do seems to be inherent in an effort to govern the quality of the environment is a very serious political issue that I think we have to face.

Now, I am going to leave you in suspense about my own approach to this, which, in the end, will come out to be rather un-Reaganish. Meanwhile, I want to put before you the simple notion that such things as bureaucracy and political power have a very important relationship to what you try to do in the name of ecology. I don't need to tell you that, after what happened with respect to Proposition 15 (the California nuclear-safeguards initiative). That was an ecological move on the part of some people. They felt that there was an unacceptable ecological assault originating from nuclear power plants, and what happened? That attempt ran into an enormously powerful economic and political machine. You were confronted with not only the economic and political power of the utilities, but of the labor unions. In other words, it became a political issue. The point I want to make, then, is that there is no way of avoiding the problem.

However, I want to talk about the problem before we discuss solutions. There is no way of avoiding the problem, but there is some kind of conflicting interaction between a desire to maintain the quality of the environment and very serious issues in our political lives having to do with freedom—having to do with who exercises power and who controls what happens in the country. What I want to do is share with you some of the facts and ideas that I think we all have to understand in order to try to cope with this very peculiar situation, and, I think, decisive interaction that is really going to govern what we can do about the environment, let alone about a number of other things.

One of the things, then, that I want to do is to establish the existence of this tension between a sensible approach to the environment and the resulting economic and political difficulties that emerge. Now, to get at that interaction (and you notice in figure 1 that I put the ecosystem on the left and the economic and political power on the right), the real question is what's in between? At what points do economics interact with the environment? At what point does political power or economic power interact with the laws of

FIGURE 1

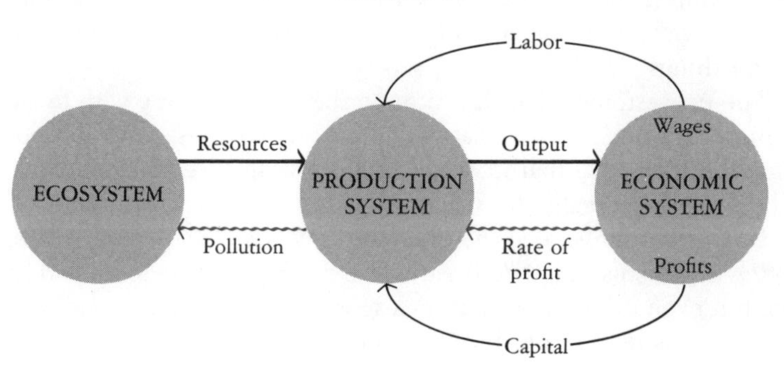

ecology? I want to make a very simple proposal to you, which largely defines what I am going to talk about. I am going to propose that there is, between these two realms of our lives—the Sierra on the one hand and Pacific Gas & Electric Company on the other—something that I am going to speak of as the production system or the way in which we use our resources to produce our goods and services. In other words, the factories, the farms, the communications system, and so on, all between the economic and ecological systems. What I am going to propose then, is that the way to think about this conflict is to look at the way in which three big systems interact: the ecosystem, the production system, and the economic system. I am going to use some tables and figures which will illustrate the interaction and then suggest some basic data that I believe will help us think about what these interactions are. Then we will come back to the question of the apparent conflict between environmental quality and political freedom. The first figure is just a diagram, in a sense. Think of these three systems as three circles. The ecosystem is what? Think of it as the thin skin of the globe. The air, the water, the soil, and what's beneath it: coal, petroleum, ores, and so on.

It has its own laws. The ecosystem was here before we were. It represents a very complex interaction between living things and the non-living environment in which living things govern the chemistry of the earth's surface. For example, the fact that there is oxygen in the air is a consequence of green plants, and the fact that we can survive with oxygen is a consequence of green plants having evolved on the earth's surface before we did. There is a complex interaction of one living organism with another. There are cycles that connect these things, and no single organism can survive by itself. We can't survive without plants, and plants can't survive without the rest of living things.

The ecosystem is that thin skin on the earth's surface with living things churning in chemical and physical activities. It is autonomous in the sense that it would do very well if we all went away. The ecosystem really doesn't need us.

But we certainly need the ecosystem. Why? Well, in part, we are living organisms in it. We breathe the oxygen, eat the food, and so on, but even more important, in a sense, is the fact that we use and produce goods. We produce the goods and services that we want, using resources all of which come from the ecosystem.

Now, I think that this is an extremely important point for everybody to understand: the ecosystem is the only place where we get the raw materials which we need to make everything that we use. For example, the oxygen in the air is essential for making steel. This, in effect, says that air is essential for making steel, which, in effect, also says that a steel plant operates by the grace of the photosynthesis of green plants. This is literally true; there is no other way to get oxygen on the earth's surface—that's how we get oxygen.

The second circle, the production system, includes the factories and the farms, which take the resources from the ecosystem and convert them in various ways making the goods and performing the services that we want. Now, the production system itself has an output of goods and services, but we don't use those directly. Usually they are converted into amounts of wealth that they represent. That's the economic system, which is money, credit, wages, profits, debts, and so on. I want to preserve the sense of the production system as something physical where cars are made, food is produced, communications are sent.

Thus, in this sense, the production system lies between the ecosystem and the economic system. The production system depends on the ecosystem and on the resources out of which it generates the wealth that is manipulated by the economic system. Now, if you think about it for a moment, you can see that there ought to be a logical set of relationships. For example, if production (the steel mill) is dependent on the ecosystem for its oxygen, then you ought to build a steel mill in such a way that it does not destroy the capability of the ecosystem to produce oxygen. In other words, it ought to be compatible with and governed by the character of the ecosystem. The production system should be molded and made to fit the ecosystem which it depends on. Now, in fact, it doesn't work that way because we build steel mills that kill plants, that produce pollution, and as the zigzag arrow (in figure 1) shows, the production system actually has an impact on the ecosystem. We build cars that produce smog and put chemicals into the air that are not good for people and plants. We build chemical factories that produce substances that are poisonous to animals, people, and plants. Sometimes we do it deliberately, sometimes accidentally. The production system is, in fact, operating in such a way as to be very heavily incompatible with the ecosystem on which it depends, and that, in essence, is the environmental crisis—that we have a production system which is destroying its own ecological base. It can't go on that way.

The question then arises, why have we so foolishly designed our production system? Logically, the economic system, the banks, depend on the factories and farms that produce the wealth. Obviously, you couldn't run banks if all the factories and farms closed down. Therefore, the economic system should be, in turn, dependent upon the production system. But that's not the way it works. There is a form of dictating the design of the production system which originates in the economic system. That's what business men call the "bottom line," or the rate of profit. Generally, the kind of factory that is built (i.e., the production system), is one that someone guesses will yield a satisfactory rate of profit. This, in turn, is often incompatible with the ecosystem.

By providing a sense of the interplay which exists here, we can see where trouble arises. Now, those looping arrows in figure 1 are quite important. The lower one says that capital has to be derived from the economic system and fed into the production system. That's a very simple but profound thing. If you want to start farming, you have to acquire enough wealth to buy the land and the machines and the seed before you can begin to create that sector of the production system. In other words, you have to acquire capital. You have to accumulate it in a big enough amount to do the thing that you want to do, so that, in effect, part of the output of the production system is withdrawn and fed back into the production system for the sake of building up more of it or replacing worn-out parts. This whole feedback of capital turns out to be the very important issue as I will show below. In the same way, labor, which is supported by wages, is essential for the production system. People must be paid so that they can live and work and operate the production system.

We must recognize that the production system lies at the interface between the ecosystem and the economic system—I don't have to argue that the economic system has a great deal to do with political power—and that if we want to think about the relationship between the two ends, we have to understand the middle. That turns out to be moderately complicated. I want to discuss a couple of examples of how the design of the production system influences the *relationships* between the ecosystem and the economic system.

The first example is oil, which is a serious political problem. The price of oil continues to rise. Now, why does it go up? It rises because the United States does not produce enough domestic oil to take care of its own needs, and that has led to the embargo and the whole series of very difficult questions. The question I want to ask is this: Why is this? Oil, gas, and coal are products of the photosynthesis of ancient plants that were laid down under the earth millions of years ago. They are what is known as a non-renewable resource, that is, there is a certain amount of oil and coal and gas, and when you use a barrel of oil, there is one less barrel left. In other words, it is a limited resource—it is non-renewable.

You have all heard that we are running out of oil. Now, this is an ecological statement. You have heard ecologists talk about the need for having renewable resources and the difficulty you run into if you use the non-renewable source. You also hear that what's going to happen is that we are going to use it all up and there won't be any left. Now, I want to make the following point and illustrate it with data. The fact that oil is non-renewable translates itself into a very profound economic process. Let me show you how it works. Figure 2 is taken from a report of the National Petroleum Council which consists of representatives of all the major petroleum companies in the United States and is an official part of the Department of the Interior. The Department of the Interior wanted to set up an educated body expert about petroleum, and naturally, it had to base itself on the people who knew about petroleum. In the United States, petroleum is looked for and produced by private companies. So this is a group of private companies essentially part of the government. They know most about petroleum in the United States. In this report they gave two figures, Curve A and Curve B. Curve A is

FIGURE 2
OIL ECONOMICS

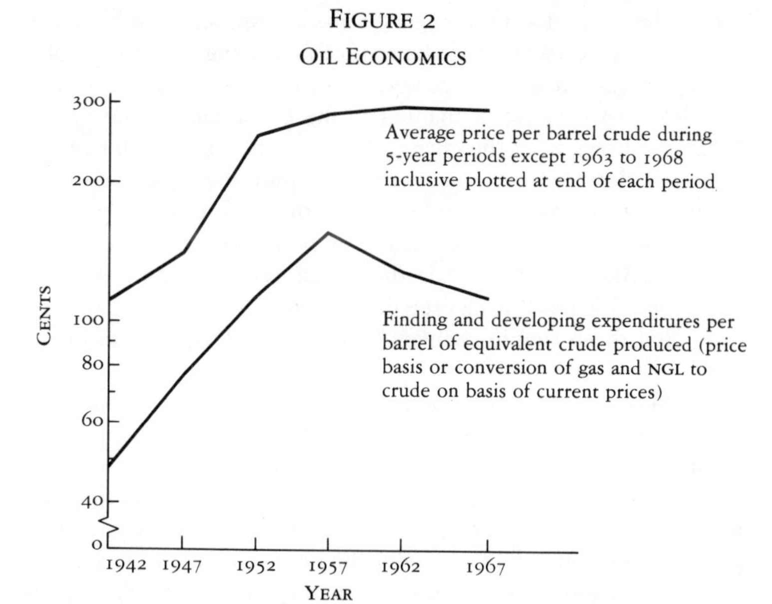

SOURCE: *Future Petroleum Provinces of the United States.* Washington, D.C.: National Petroleum Council, 1970.

the average price of domestic oil running from 1942 to 1968. You will notice that the price went up and then leveled off. In other words, beginning around 1952, the price of oil was relatively constant. This is affected by inflation, and, as I'll show you in a moment, actually, the uninflated price of oil went down. It is a basic fact that the price of oil leveled off. Curve B is the relative expenditures for finding oil per barrel. In other words, how hard did the companies have to work—how much did they have to spend to find oil? You'll notice that it went up very sharply, and then, in 1957, it dropped. The reason why it dropped was that they quit looking; the number of exploratory wells drilled began to drop off. The explanation given in the report was that the effort to find oil became unprofitable because it became so increasingly expensive to find. At that point, incidentally, the oil companies went abroad and produced oil in Venezuela and the Mideast.

What I am saying is the rising cost of oil is the expected result of what you might call an ecological fact. Oil is limited in amount, so that every time you produce a barrel, the next one is harder to get out of the ground—it's deeper, it's less accessible. There's going to be a constant escalation in the cost of producing oil because of an ecological fact: it is a non-renewable resource. But this, as you can see, leads to very rapid changes in the economics of the system. Figure 3 shows what the changes are. This figure is again from the National Petroleum Council. In 1969 they did a very elaborate study of what was going to happen to the price of oil. The report was published in 1971, before the embargo. Before the embargo, the solid line on the left is the actual price of domestic oil in uninflated dollars. You can see that it goes down slightly. Then they predicted three simple changes in the price, all of them going up. Those three lines are based on three assumptions. The lower line is the assumption that the companies would be satisfied with a 10 percent profit, the middle one, that they would accept a 15 percent profit, and the upper one, a 20 percent profit. In other words (and you notice that the 15 percent line pretty much matches the historic line), what that says is that in order to maintain a 15 percent profit in domestic oil production, the price would have to begin to rise in 1973 and thereafter go up in an exponential, escalating way. This is an accurate description of what we know about the ecology of oil. If it is true that it is limited in amount, then every time you

FIGURE 3

PROJECTED (NATIONAL PETROLEUM COUNCIL) AND ACTUAL CRUDE OIL PRICES (1970 Dollars)

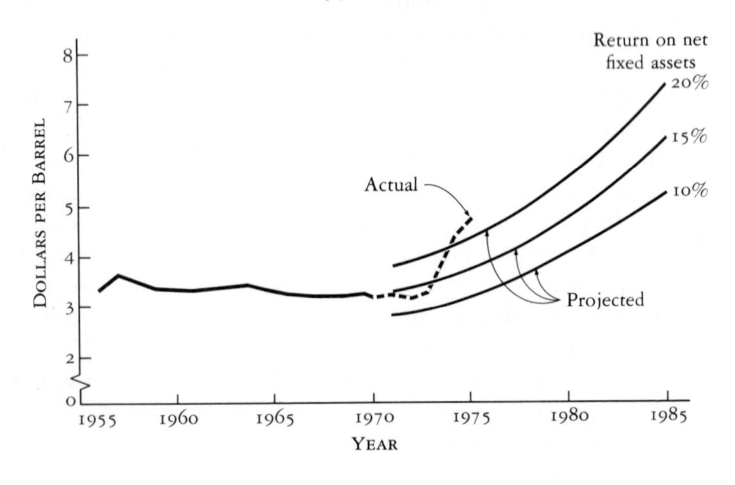

take some out, it becomes increasingly expensive. They did a computation and figured out what the price would have to be. Incidentally, I plotted on their graph the dotted line which is the actual price. What has happened is that it started out up at the 15 percent curve and it has now jumped to the 20 percent curve.

This is a very interesting case of a self-fulfilling prophecy. The oil industry prophesized, even though the price had remained constant for twenty-five years, that in 1973 the price of oil would have to rise. It did. I don't say that this means they conspired with OPEC—nothing of the sort—it's exactly the other way around. What it means is that the oil ministers of OPEC also went to the Harvard Business School, and they knew perfectly well that the American oil industry had to have a higher price for domestic oil. (Mideast oil is much cheaper to produce, but why should they sell it at a cheap price when the alternative to the American industry was to buy higher-priced domestic oil—so they raised the price.)

In other words, the rise in the price induced by the OPEC countries was nothing but a foreshadowing of what I will call a natural, or if you like, ecological process translated into economics. A lot of people talk about non-renewable resources and what it means ecologically. This can be translated into some obvious ecological examples. Take the passenger pigeon. The passenger pigeon is extinct. It is a classical case of a threatened species that didn't make it. In 1800 it was estimated that there were several billion passenger pigeons in the United States. They turned out to be rather stupid birds; you could walk up to them and hit them with a stick and they would die. They were also unfortunate in that they were rather tasty. As people began to fill the continent, they simply killed all the passenger pigeons, and by 1900, I think, the last pair disappeared. We killed the species off.

It turns out that there are data on the price of passenger pigeons since they were a market item. In 1800 you could buy a passenger pigeon for a penny. In 1900, just before they disappeared, they cost 20 cents each. What I am trying to say about the business of extinction is that when you wipe out a non-renewable resource the process is translated into an escalating price.*

Another example is the buffalo. Between 1870 and 1880 we

*Even though these pigeons were a renewable resource, they became non-renewable as they were killed off fast enough.

essentially wiped out the large herds of buffalo. In those ten years, the price of buffalo skins rose sixfold. We're seeing exactly the same thing here. Using up a resource means an escalation in price. There's a big difference between passenger pigeons, buffalo skins, and petroleum—there is no substitute for energy, while there is a substitute for buffalo skins.

Again, in looking at ecology and economics we're talking about a well-known ecological process. And instead of this process carrying itself through only a physical process, it is very quickly translated into an economic process—an escalating price. As long as you have energy coming from a non-renewable source (oil, natural gas, coal), eventually its price will have to rise as it gets used up. Figure 4 is a diagram which reminds me of when I wrote *The Poverty of Power.* I very carefully looked into the estimates of oil reserves from the point of view of the economic assumptions that we used to make the estimates. I came up with a figure that was about 450 billion barrels of total oil resources in the United States, whereas most geologists were talking about 150 billion barrels. I came up with that figure because I discovered that the estimates made by the geologists were based on the current price of oil. When my book came out, the Geological Survey was a little upset. The head of the division that handles all this wrote me a long letter and sent me a copy of a report which, he said, proved that I was wrong. The report, USGS Report No. 725, says right on the front page that they have studied the reserves of oil and that there are roughly 150 billion barrels. I read on. It said that this was based on certain assumptions. One of these assumptions used the 1974 price of oil. I read further, and toward the back of the report was figure 4. These are the various ways of measuring reserves—you add the numbers going to the right and you get the total amount. There's a line here that says "economic" and one that says "sub-economic." Their 150 billion barrel figure is the upper numbers added up. That is, the oil we have, which is economically recoverable at 1974 prices. Where is it written in golden tablets that the people of the United States will take oil out of the ground only when it is economically suitable for the oil companies of the United States? If you're a geologist, you ought to be talking about oil, not oil companies' oil. The USGS felt that way and they were a little nervous about it too, so they put in the bottom part called "sub-economic reserves." If you add it all up, it comes up to my figure, or 450 billion barrels. In other words,

FIGURE 4
CRUDE OIL AND NATURAL GAS RESOURCES
OF THE UNITED STATES

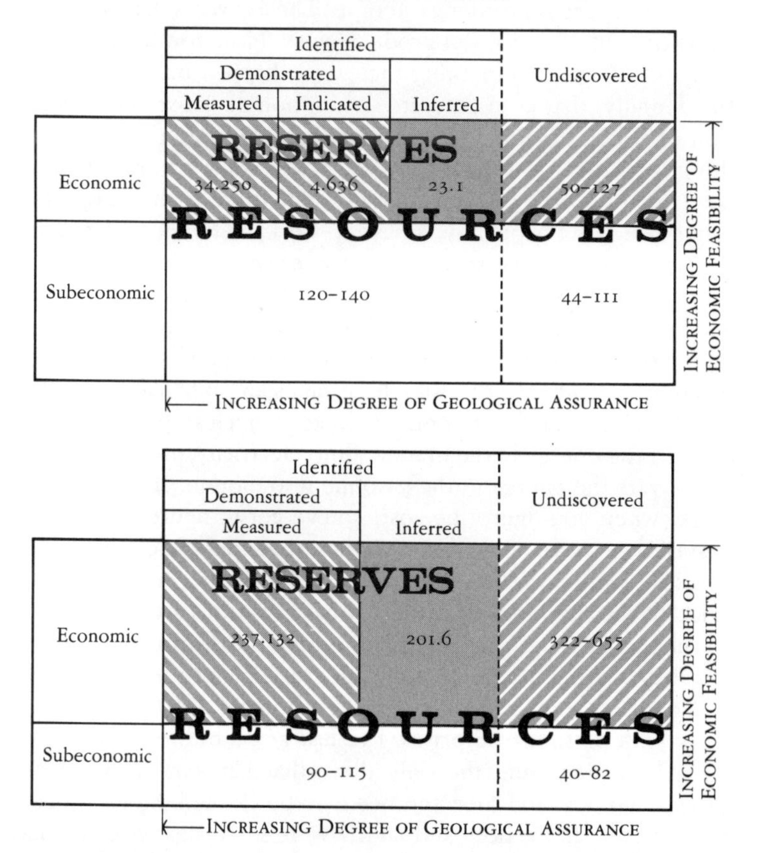

Top: Crude oil resources in billions of barrels; total United States cumulative oil production, 106 billion barrels (31 December 1974).
Bottom: Natural gas resources in trillions of cubic feet; total United States cumulative gas production, 481 trillion cubic feet (31 December 1974).

they sent me a report that actually proved what I was saying. The economic assumptions determine the available oil. This is just to remind you, then, that you cannot isolate what's happening in the ecosystem from the economic assumptions.

Let me now very briefly talk about the science of energy called thermodynamics. As you know, there are two laws of thermodynamics. The first law is really idiotic. It says that energy can neither be destroyed nor created; in which case, what's the fuss about? Why are we running out of it? The answer is that energy has value only insofar as it can produce work. Now you will ask: What is work? Work is something that won't happen unless you do it. Incidentally, that is an accurate description of the scientific definition of work. It's literally true. For example, if you don't do anything to water in a waterfall, it falls. If you want it to rise, you will have to do work. So, if you want something to happen, which otherwise won't happen, you have to work, and to do that requires energy. Now, the reason we need energy is to do the work required in the production system—to make the wheels turn around and to keep a building warm. Therefore, we have a problem of taking the source of the energy and converting it to work. It takes a machine to do that. For example, it takes an oil burner to convert the oil into heat for a room; it takes a nuclear power plant to convert the energy in the uranium into electricity, and it takes a car to convert the energy in the gasoline into motion. If you want the work, when you figure the cost, you've got to figure not only the cost of the source (and we've just described that the cost will go up if it's an unrenewable source), you also have to ask: What's the price of the machine? Machines come in various styles. Let's take an oil burner. That's an essential machine for getting the work out of oil to warm a home. I would call that a mature technology.

What do we mean by a mature technology? It's one which will work with no future surprises. I've had an oil burner in our house for a long time, and the only thing that happens is that it gets clogged up now and then and needs to be cleaned. So far, nobody has knocked on the door and said, "We're sorry, next year you have to put a big radiation-proof shell around this thing, because we've just discovered it's bad for the environment not to have that shell." In other words, there are no surprises. It is a mature technology. Again, how do I know it? The price remains constant. If you look at the price of oil burners, it's gone up with inflation, but that's it.

Therefore, I want to suggest that we have to look at the machines for producing energy and ask: Are they technologically mature or immature? An example of a technologically immature machine is a nuclear reactor. First, I want to show you the *source* of the problem. Figure 5 shows the predicted prices of the uranium fuel, and the

FIGURE 5

RANGE OF COST-BASED PRICES FOR URANIUM SUPPLIES, 1980-2000

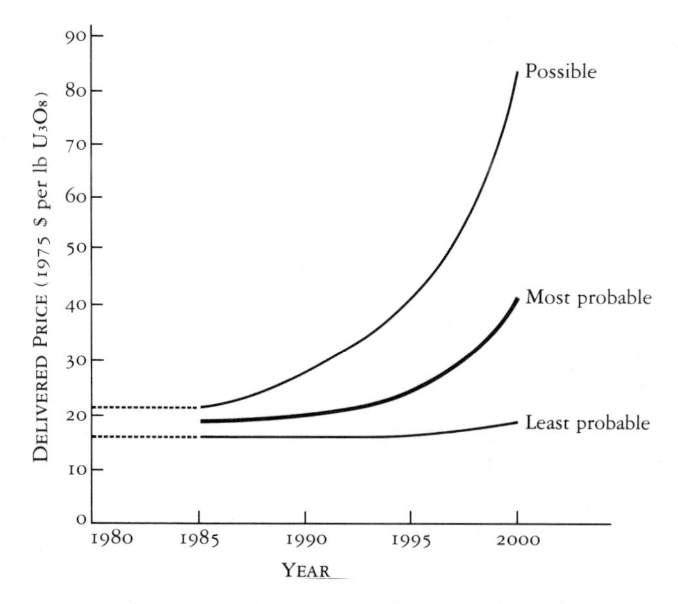

The three scenarios are based on three different assumptions concerning the purity of uranium ores yet to be discovered. While this figure shows the expected trend in uranium prices, it does not take into consideration the rapid increase in uranium price in 1975, when the price rose to over $40 per pound.

SOURCE: Edison Electric Institute, *Nuclear Fuels Supply*, New York, 1976.

same old escalating curve, because it's a limited resource (a limited amount of uranium). When you produce uranium, you take the cheapest, easiest-to-get-at fuel first, then you take the more expensive fuel, and the price goes up. Actually the figure is outmoded since the price is well above this prediction already.

Now let's talk about the machine. In the first place we can ask: What about the price of the electricity? Again, I am using price as evidence of how hard it is to do it. Figure 6 shows the results of an official report of the average price of electricity in the United States from different energy sources. You will notice that the most expensive is oil because the price of oil has gone up. The next expensive is nuclear power, and its chief competitor (because oil is now being withdrawn from the production of electricity), coal, is significantly

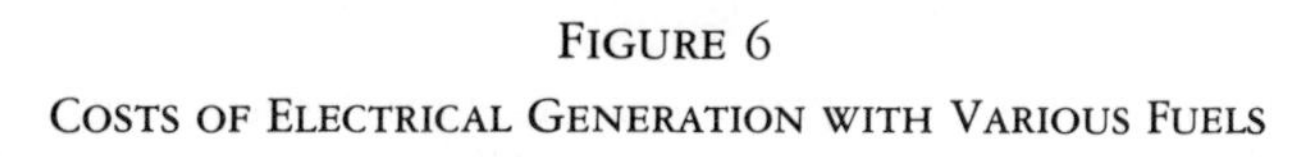
FIGURE 6

COSTS OF ELECTRICAL GENERATION WITH VARIOUS FUELS

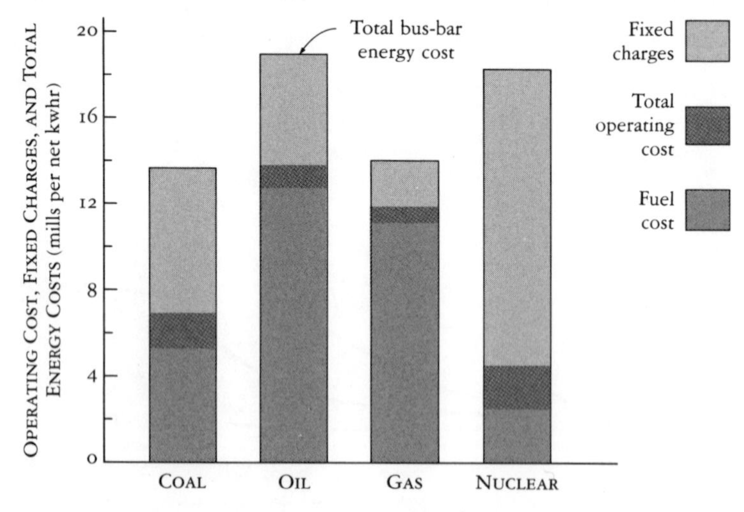

SOURCE: 19th Steam Station Cost Survey, *Electrical World,* 15 November 1975.

lower in its cost. In other words, coal is cheaper on the average. (In certain parts of the country it is not. For example in New England where coal is rather expensive, nuclear power is still a little cheaper.) This is a surprise because, when the utilities were persuaded to build nuclear power plants, they were told that nuclear plant electricity would be cheaper than coal-fired electricity. Nuclear power turned out to be an immature technology.

Figure 7 shows the cost of electricity produced by nuclear power plants divided into three categories. They are the cost of fuel, operation and maintenance expenditures, and the capital cost of building the nuclear power plant. In 1966 half the cost of the electricity was represented by the cost of the machine. In 1975 it rose to 77 percent. In other words, this was an unexpected jump in the cost of the machine needed to produce this energy. Why? Figure 8 is from a study done by I. C. Bupp at the Harvard Business School of the increase in the cost of power plants. What do we mean by the increase? When you bid for a nuclear power plant, there's a certain price given; when you actually pay for it, it's a higher price, and each year it goes up. Figure 8 shows the actual data for the price, from year to year, of nuclear power plants, coal-

FIGURE 7

PROJECTED COSTS OF NUCLEAR GENERATION OF ELECTRICITY

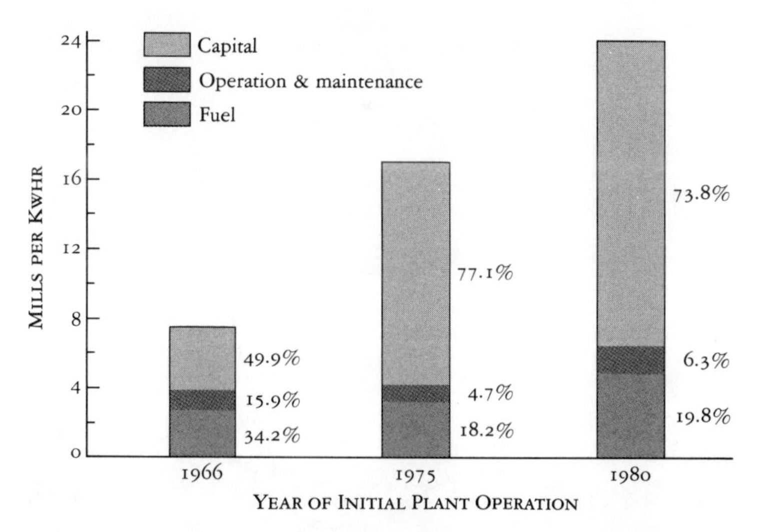

SOURCE: R. E. Scott, "Projections of the Cost of Generating Electricity in Nuclear and Coal Fired Power Plants." St. Louis: Center for the Biology of Natural Systems, Washington University, December 1975.

fired plants, and refineries, starting with 1970 at 100. What you see is that nuclear power plants are rising in cost three times faster than the cost of a comparable coal-fired power plant.

Bupp did a very careful statistical study of the various nuclear power plants, relating various factors to the rise in their cost. What he found was a good statistical correlation between the rate of increase in the cost of nuclear power and the duration of the hearing held to license the nuclear power plant. People in California know that's exactly what the utility people say. And of course, those damned environmentalists come in, delay the hearings, and all this time we have to pay interest on the capital and prices go up because of inflation. The delay in construction, this expenditure of time, is responsible for the rapid rise in the capital cost of a nuclear power plant.

Time affects a nuclear power plant project in two ways. Once is during the hearings and the other effect occurs during the construction. Bupp did a statistical analysis of the relationship between the increase in cost and the duration of the construction period, and he

FIGURE 8

COST INDEX TREND FOR NUCLEAR PLANTS, COAL PLANTS. AND REFINERIES

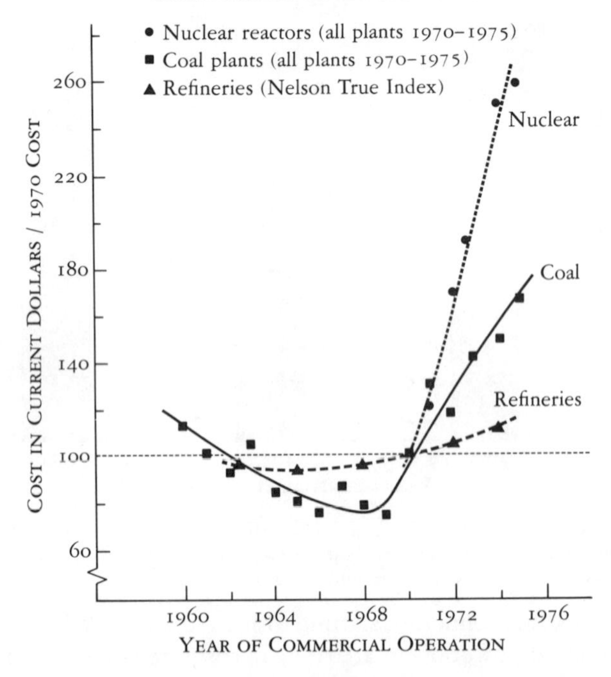

SOURCE: I. C. Bupp et al., *Trends in Light Water Reactor Capital Costs in the United States.* Cambridge: Center for Policy Alternatives at the Massachusetts Institute of Technology, 18 December 1974.

found a poor correlation. What he said was that the first correlation must be a proxy for something else. What is it a proxy for? If you look at the records of the hearings, you see what it is. Sure, the environmentalists came in and pointed out, as many did in California at the Bodega Bay reactor years back, and told the AEC that this reactor is sitting on the San Andreas Fault. It took a little time to point that out, and after a while, the AEC acknowledged that by gosh, it is on the San Andreas Fault, and it *would* crack the containment shell open if there were an earthquake. We'd better not do it and Pacific Gas & Electric had a four-million-dollar hole in the ground, which was finally used for some other purpose. That is a perfect example of how the introduction of an environmental concern raises the capital cost of nuclear power plants. That was four million dollars that produced nothing. What Bupp said was that about 65 percent of the increased cost was a result of conditions of that sort being introduced in the hearings, as well as such things as having to have an emergency cooling system for the core, making changes in certain piping systems, and so on. These increased the cost. In effect, the nuclear power plants were being redesigned in hearings and that's not an efficient way to build a machine. It literally is not, and incidentally, this has not stopped. In the October 21, 1976 edition of the *New York Times,* it was said that three engineers in the Nuclear Regulatory Commission announced that the Nuclear Regulatory Commission had suppressed a series of faults concerning safety problems they had pointed out in nuclear power plants. They said they were suppressed because the Commission decided it would raise economic difficulties for the utilities.

What this means is that the machine has not yet matured. I'm making a very simple point. I don't want to argue the question of radiation dangers. All I am saying is that most of the rising cost of nuclear power plants is a result of newly accepted responsibilities for safety and environmental controls. You have to ask yourself why this should be. Why isn't it as well developed as a steam engine or an oil burner?

The answer, again, comes from thermodynamics. One of the things about thermodynamics is that it tells you the efficient way to carry out a task that requires work is to get a good match between the character of the energy and the character of the task. Let me give you a simple illustration. If there is a fly on the wall and it needs to be killed, that's a task, and it can be accomplished by energy administered through a fly swatter. That is a moderately efficient way to kill a fly. Now, I can also kill that fly on the wall by aiming a cannonball at it, and that will kill the fly but at great expense to extraneous things, like the wall. In other words, the cannonball is too powerful for the task.

Now, what is the task in the nuclear power plant? The task is to boil water. That's it, nothing fancy—boil water to make steam and the steam drives the generator. To run the generator efficiently you need to make steam with an input temperature of roughly 1,000° or perhaps 2,000° F. What you need is a source of energy that will operate at 1,000° to 2,000°. An oil burner will do that, a coal-fired plant will do that, But what about a nuclear plant? The inherent temperature of a nuclear power plant—inside where the reaction takes place—is the temperature equivalent of somewhere between 1,000,000° and 10,000,000°. Some of that energy is translated into radiation, into the cracking of pipes, into "the wall falling down," for example, and you have to work to contain all of that excess energy. What I am saying is this type of system is immature because it is a thermodynamic mismatch. The consequences of this is that we are dealing with a way of producing energy which is bound to increase in price.

Let me tell you about another aspect of the energy problem. The United States now has an energy plan. You might not know it, but ERDA published such a plan about six months ago.* I testified before the Senate on the plan, and to do that I went through everything they proposed to do in the next five, ten, and fifty years. I examined them from the point of view that I have just gone through with you here. Are their plans based on a non-renewable resource so the price will escalate? Or are they based on a renewable source? Are they based on a mature technology or an immature technology? I'll give you the short answer which I said before the Senate. If what they proposed to do were followed-out, the price of energy would rise for the next 75 years. First, they were going to start intensive production of oil which means the price goes up. Second, they wanted to convert coal to oil, which as I'll show you in a moment is very costly. Third, they wanted to build more nuclear power plants, including the breeder, and if the current nuclear power plant is immature, the breeder is infantile. Then they said that maybe we would also have fusion, which, of course, is prenatal.

Table 1 shows the capital cost of different ways of producing energy. The top one is the one I have already talked about. The amount of energy you get per dollar of capital invested will fall about fourfold in the next ten years or so in producing oil. If you want to produce shale oil, synthetic fuel from coal, coal gasification, all the things that the ERDA plan emphasizes, you get very little energy out of a dollar of capital invested. Thus, this is a capital-consuming approach, and it means that the price goes up.

Now, I want to show you something about the historical productivity of capital invested in energy (figure 9). What does this figure mean? It means that less and less energy is being produced per

---

*The Energy Research and Development Administration, now part of the Department of Energy.

TABLE I

CAPITAL PRODUCTIVITY OF ALTERNATIVE ENERGY SOURCES

Energy Source	Capital Productivity (BTU's per year per dollar of capital invested)
Crude Oil Production[1]	
1974 (actual)	16,800,000
1988 (projected)	4,480,000
Coal (strip mined)[2]	2,000,000
Shale oil production[3]	420,000
Synthetic fuel from coal (liquid)	254,000
Coal gasification[3]	160,000
Coal-fired electricity generation ($800/kw)[4]	28,683
Nuclear electricity generation ($1,000/kw)[4]	22,423

[1] The capital productivity of oil production was derived from information in *Oil: Possible Levels of Future Production,* Final Task Force Report, Project Independence, FEA, Washington, 1974, pp. IV-2 and IV-21.
[2] The capital investment required to produce one ton of coal was obtained from *U.S. Energy Outlook: Coal Availability,* Washington: National Petroleum Council, 1973, p. 38.
[3] The capital investment required to produce different synthetic fuels was obtained from the *Project Independence Task Force Report on Synthetic Fuels from Coal,* p. 35, and also the *Task Force Report on Oil Shale,* p. 65. FEA, U.S. Department of the Interior, Washington, November, 1974.
[4] The estimates for coal-fired and nuclear power plants are for base load power generation, operating at 75% of capacity for 1 year.

dollar invested over time. Capital is diverted from the output of the production system in order to produce energy. This implies that the efficiency with which we use this precious capital to produce energy, is falling. We are turning more and more of our capital into a process of getting energy out and we are doing it less efficiently.

Now let's look at the price. Figure 10 comes from standard figures from the Department of Commerce (the Survey of Current Business). Since 1811 the Department of Commerce has published a series of price indices for fuel and for commodities in general. This figure shows the fuel and power index divided by the commodity index, or the price of energy relative to the price of everything else. Beginning in 1930 the price was relatively high and then dropped from 105 percent of the commodity price to about 77 percent in 1950. From 1950 on it dropped very slightly, but you notice how steady it was. In other words, the development of in-

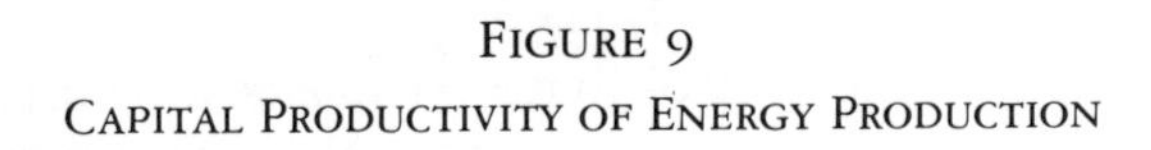

FIGURE 9

CAPITAL PRODUCTIVITY OF ENERGY PRODUCTION

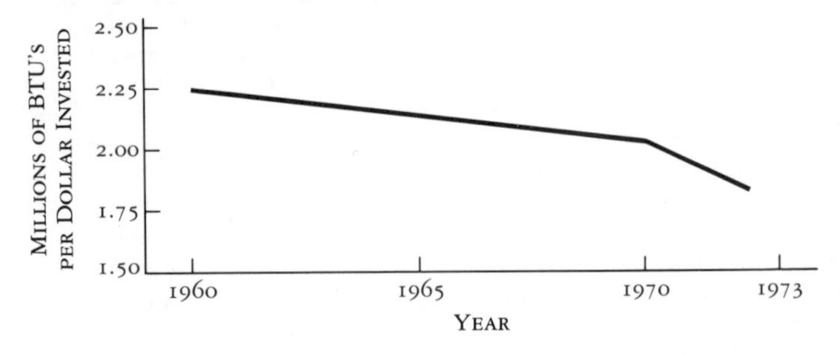

SOURCE: Data computed from capital estimates made in B. Bosworth et al., *Capital Needs in the Seventies*, Washington, D.C.: Brookings Institution, 1975, pp. 27–29: and from domestic energy production as reported in U.S. Department of the Interior, *Energy Perspectives*, Washington, D.C.: U.S. Government Printing Office, 1975.

dustry in the United States—and particularly agriculture—was assured of at least a steady price of energy and possibly a falling price.

You need energy to run a machine. You need a machine to produce goods. When you invest in a machine, you must figure the cost of running that machine, the labor, and what you are going to get for the goods that you produce. And, you have got to balance out. If you can be *sure* of the price of energy, that's a very important piece of information. What this says is that during the entire period of the development of modern industrial and agricultural technology in the United States, we have had the benefit of either a steady or a declining price of energy. Why is that important? Because unlike the passenger pigeon or the buffalo robe, there is no substitute for energy.

The economist will tell you that if the price goes up, you will substitute, but there is no substitute for energy. Just look at those points after 1973 on figure 10. This is an unprecedented event in the entire history of the United States. The price of energy relative to other prices is rising sharply and continuously, and it will go up further. At the very least, it seems to me that no politician would want to say, "Well, the price of oil is high, we know; the price of gasoline is high; but you realize that it was always too low and we were using too much and it was underpriced—so it's O.K. that the

FIGURE 10

RELATIVE PRICE OF ENERGY, 1900–1976
(Ratio of Fuel and Power Wholesale Price Index
to All Commodities Wholesale Price Index)

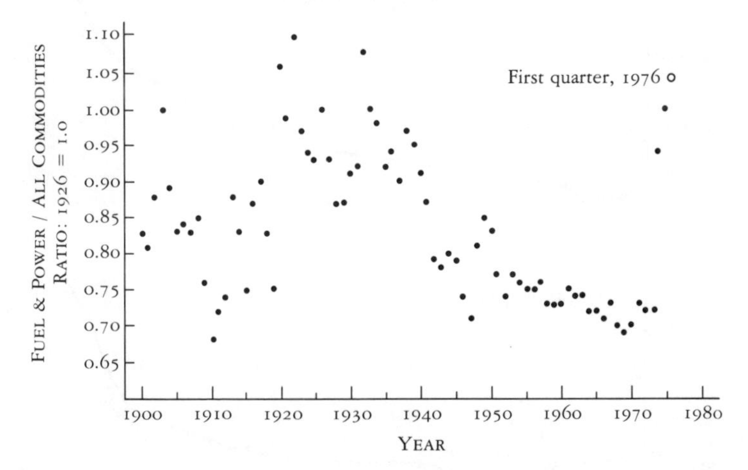

SOURCE: U.S. Department of Commerce, *Survey of Current Business,* Washington, D.C., various years.

price is high." They are talking about an absolutely unprecedented event in the history of the country as though it were like drinking a cup of tea.

At least you want to ask the following question: If the price of energy is going up, and there's no way of substituting for it and everyone uses energy, what's going to happen to the price of goods as a result of this sudden sharp rise in the price of energy?

Figure 11 shows the energy index, the wholesale price index, and the consumer price index in recent years. What you see is the steady rate of inflation until the price of energy suddenly goes up, then everything else goes up with it. In other words, the rapid inflation in the United States today has "some" relationship to the fact that we have an unprecedented escalation in the relative price of energy. How our beloved presidential candidates can speak of inflation with one hand and blithely accept the price of energy with the other hand, I just don't know. At least it has to be discussed.

The upshot of rising energy prices is inflation, a problem that is very deeply embedded in the everyday politics of life. There is no

FIGURE 11

RELATION OF CONSUMER, WHOLESALE, AND ENERGY PRICE INDICES

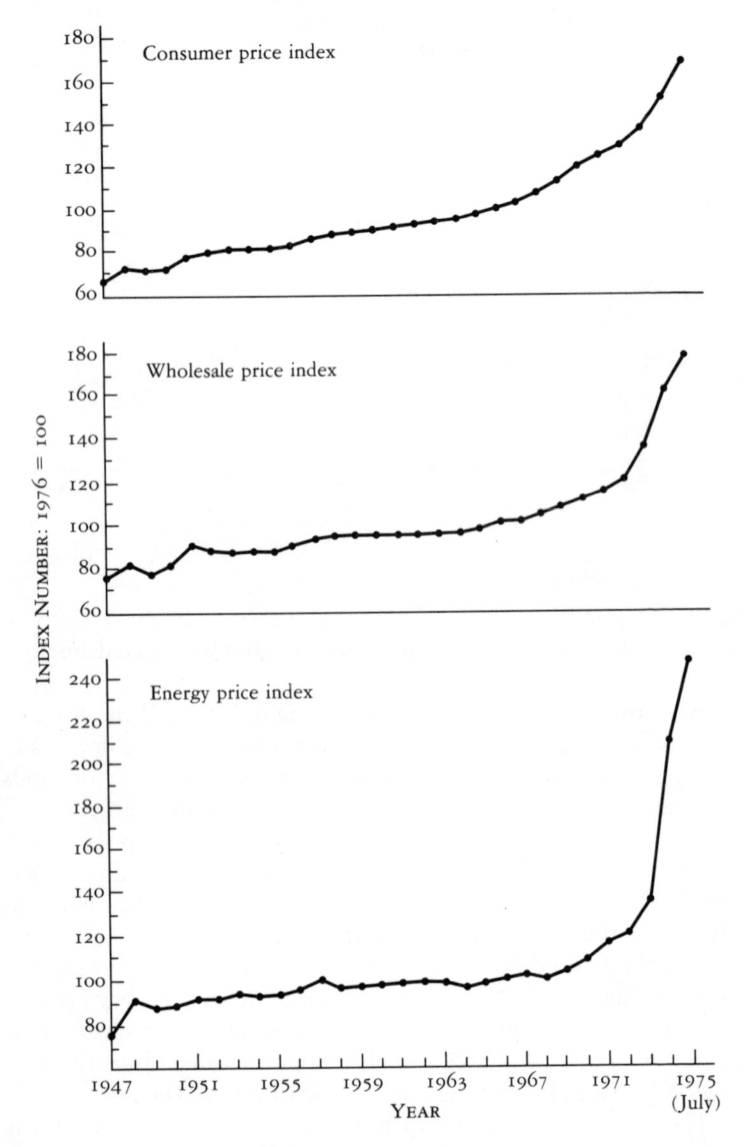

SOURCE: *Survey of Current Business.* "Business Statistics 1974" (Annual Edition) and edition for December 1975.

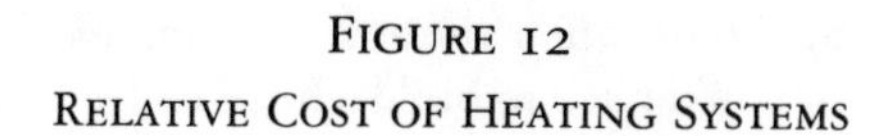

FIGURE 12

RELATIVE COST OF HEATING SYSTEMS

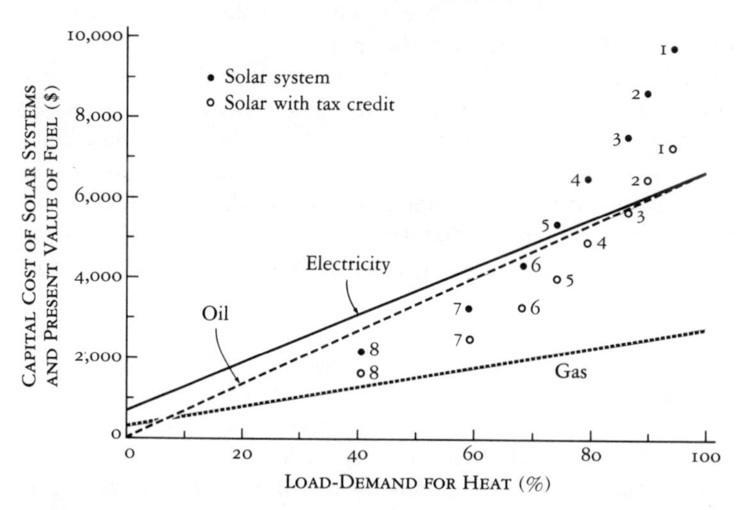

Fuel costs represent present value in 15 years assuming cost inflation of 10 percent per year; solar system cost includes capital cost at effective interest rate of 8 percent for 15 years. All data are for the climate of St. Louis.

way in my mind of thinking about our inflation problems without understanding the non-renewability of oil and the thermodynamic immaturity of nuclear power.

Solar energy is an alternative which I will mention briefly here. Figure 12 shows that it is now possible, in most parts of the country, to save money on your hot water and space heat bill by building a solar system for partial use (usually about fifty percent), and borrowing the money at eight percent effective interest and paying it back in fifteen years.* Everybody says solar collectors are terribly expensive, and they are. But, they are a hedge against inflation. Figure 12 is simply the cost of a solar system that you would need to replace given amounts of electric, oil, and gas heat in St. Louis. If you wanted to have 100 percent of your heat coming from the solar system the price is much higher, but if you are willing to make 50 percent of your energy solar and the rest electricity, your total cost will fall below that of a total electric system. You'd save about ten percent on your bill. The reason for this is very simple. There are a

*Based on a ten percent annual increase in fuel costs.

few weeks in St. Louis when there is no sun, and if you have no auxiliary system, you'd have to build a huge collector and a big tank just to get through those weeks. The sensible thing to do is use your other backup system at that point.

I will give you a simple computation on what solar space heat and water systems could do on a national level. Mr. Ford introduced a bill in 1976 for 100 billion dollars to put public funds into technologies that will raise the price of energy: intensified production of petroleum, shale oil, synthetic fuel from coal, and nuclear power. Accepting his judgment that we have a hundred billion dollars of public funds to invest in energy, we did a slightly different computation at the Center.* We took the 100 billion dollars and said that we will loan them to homeowners and owners of commercial buildings who want to have 60 percent solar energy for space heat and hot water. We will loan it to them at no interest, but they will have to pay back to the government the savings in fuel cost, but at a fixed price of fuel, and each year you pay back whatever you saved.

What will that do for the homeowner? It will stabilize 60 percent of his fuel cost, which could actually be rising, and we will get back that money and return it to the loan fund. After ten years you have the following situation. About three-fourths of the single-family homes and commercial buildings in the United States would be 60 percent on solar energy for space heat and hot water. Twenty percent of the imported oil would no longer be needed, and in the tenth year and thereafter, the government would be getting 6½ billion dollars of interest in the savings, so that in fifteen years the whole hundred billion could be returned to the government.

You will say that's magic, but it's very simple. When you have an escalating price of fuel, any device which is built that saves fuel becomes more valuable every year.

Now let's go to the business of the use of energy. I'll show you we are getting into more conflicts between the economic, production, and ecological systems. I want to talk about farming for a moment. Figure 13 shows the average yield in bushels per acre of corn (from 1880 to 1970) in the United States. You notice that from 1950 on there was a big jump, and we now know why. It was mostly the introduction of new varieties that could use large

*Center for the Biology of Natural Systems, Washington University, St. Louis, Missouri.

FIGURE 13

UNITED STATES CORN PRODUCTION, 1870–1970

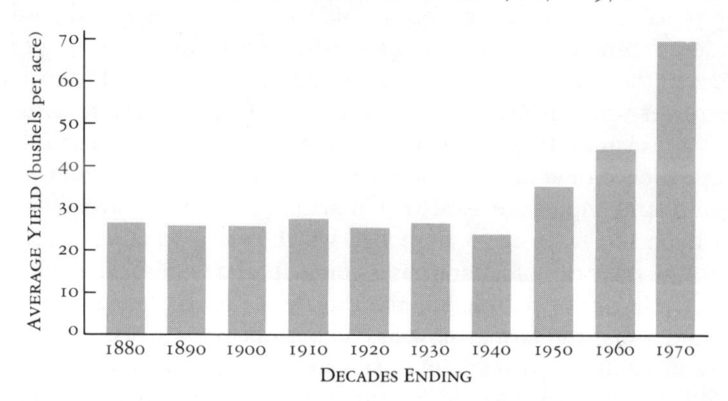

SOURCE: Keith C. Barrons, *The Food in Your Future,* New York: Van Nostrand Reinhold, 1975.

FIGURE 14

ENERGY INPUTS IN CORN PRODUCTION, 1945–1970

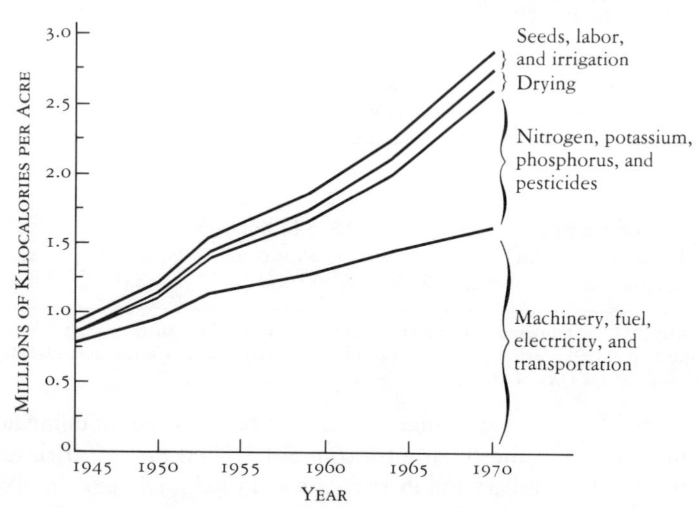

SOURCE: David Pimentel et al., "Food Production and the Energy Crisis," *Science,* vol. 182 (1973), p. 443. Cited as reprinted in P. H. Abelson, ed., *Energy: Use, Conservation and Supply,* Washington, D.C.: American Association for the Advancement of Science, 1974, p. 41.

amounts of fertilizer. What about the energy that goes into producing corn? Figure 14 shows the total energy going into corn from 1945 to 1970, and this big expanding bar is the fertilizer and pesticides. Figure 14 presents the national averages. In Illinois 47 percent of the energy required to grow corn is in the form of nitrogen fertilizer which is made out of natural gas. Only 18 percent is fuel to run tractors. In other words, what we've done is to expand the dependence of agriculture on energy inputs. Now, I have told you that energy inputs are escalating in price. So, what we have done is to hitch the farm up to ammonia, which is skyrocketing in price because the price of natural gas is going up, and to pesticides, which are made out of petroleum, and their price is also going up. The result is that the farmer now has higher operating costs.

You might say that's fine; obviously he is getting more income. Well, the data disprove this notion. Figure 15 is the total economic budget of United States agriculture from 1950 to 1972. The top line is total farm income. You notice the difference between total cost and total income, the net income that supports the farmer and his family. I am saying a very simple thing that most farmers know. Yes, they have a bigger income, but their expenses have gone up to match.

TABLE 2

FARM INCOME COMPARED TO MEDIAN FAMILY INCOME
(Constant 1967 Dollars)

	1950	1970	Percent change
Total farm income	$18 billion	$13 billion	−28
Income per farm	$3200	$4600	+46
Median family income in U.S.	$4604	$8484	+84

SOURCE: U.S. Department of Agriculture, *Agricultural Statistics*–1972 (1974 volume also used), Washington, 1972; and U.S. Department of Commerce, *Statistical Abstract of the United States*, 1975, Washington, 1975.

Table 2 shows this change in terms of real income. In uninflated dollars in 1950 the total net income of United States agriculture was 18 billion dollars and in 1970 it was 13 billion dollars. In other

FIGURE 15 (*on facing page*): GROSS AND NET INCOME FROM FARMING, 1950–1973.

SOURCE: U.S. Department of Agriculture, *Agricultural Statistics*—1972 (1974 volume also used). Washington, D.C.: U.S. Government Printing Office, 1972.

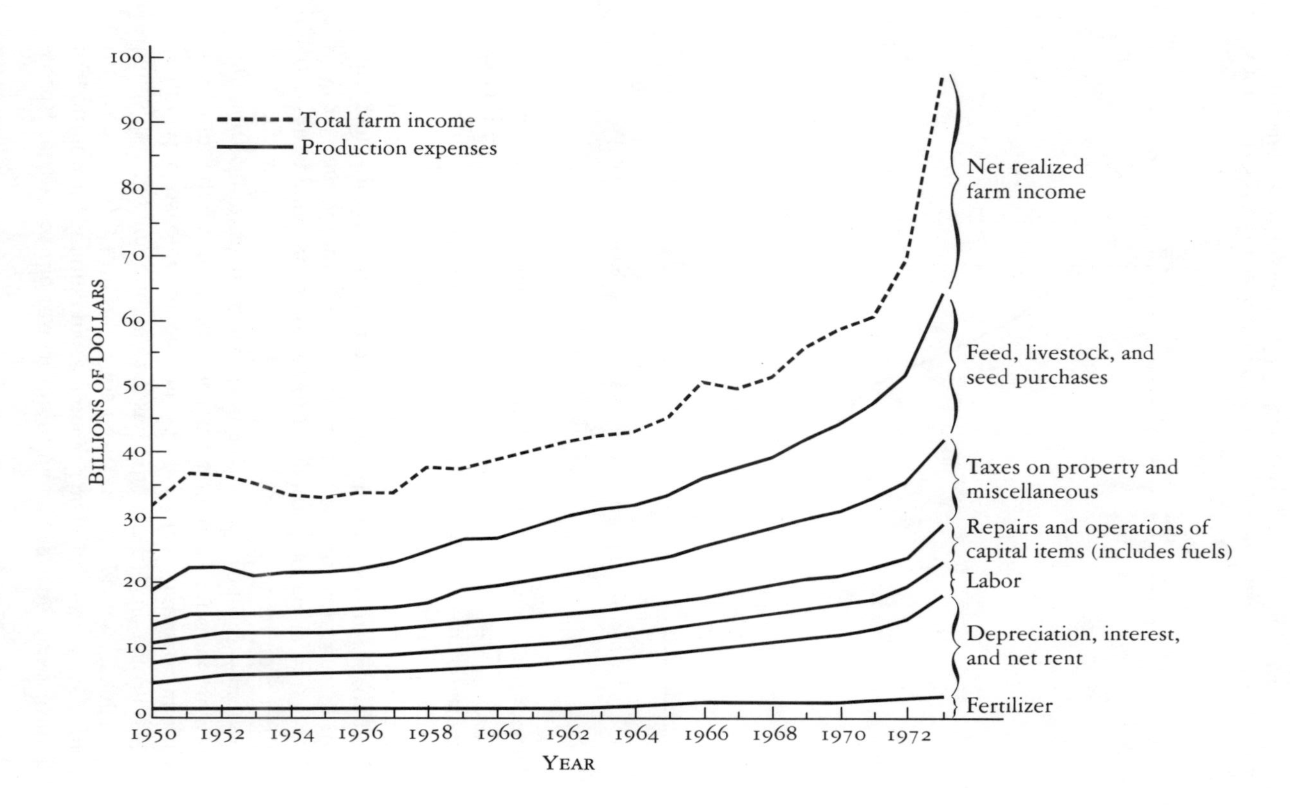
100
90
80
70
60
50
40
30
20
10
0
Billions of Dollars
Total farm income
Production expenses
Net realized farm income
Feed, livestock, and seed purchases
Taxes on property and miscellaneous
Repairs and operations of capital items (includes fuels)
Labor
Depreciation, interest, and net rent
Fertilizer
1950
1952
1954
1956
1958
1960
1962
1964
1966
1968
1970
1972
Year

words, relative to the rest of the economy, agriculture was suffering, but it was spending more money. The farmers we work with at the Center say, "Oh yes, a lot more money passes through our hands, but the amount we earn stays the same." The reason is that farming has become dependent on inputs from industry, and it is now competing with the petrochemical industry for something else the petrochemical industry itself uses and produces—ammonia, propane, and other chemicals. The result is that farming has become an appendage, a colony of the petrochemical industry.

What I am pointing out is that instead of using solar energy to fix nitrogen from the air, you buy it from the petrochemical company. The results in the farm now become susceptible to each economic disturbance that occurs in the petrochemical industry. Figure 16 shows that you need seven times as much capital to run a farm now as you did in 1950.

Figure 17 shows another example, where industry is replacing

FIGURE 16

PRODUCTION ASSETS PER FARM WORKER, 1950–1972

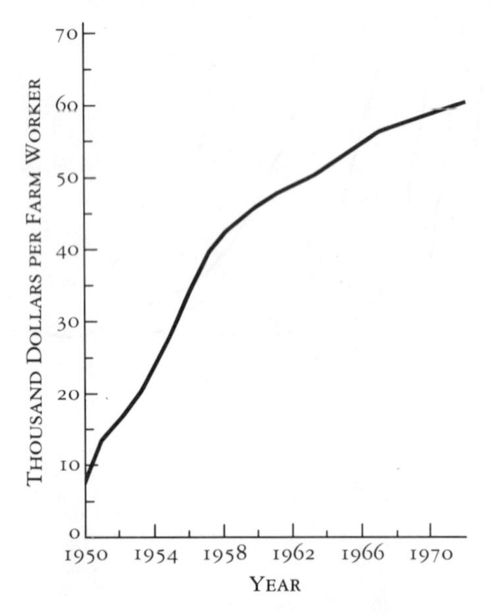

SOURCE: U.S. Department of Agriculture, *Agricultural Statistics*—1972 (1974 volume also used). Washington, D.C.: U.S. Government Printing Office, 1972.

FIGURE 17
REPLACEMENT OF SOAP BY DETERGENTS

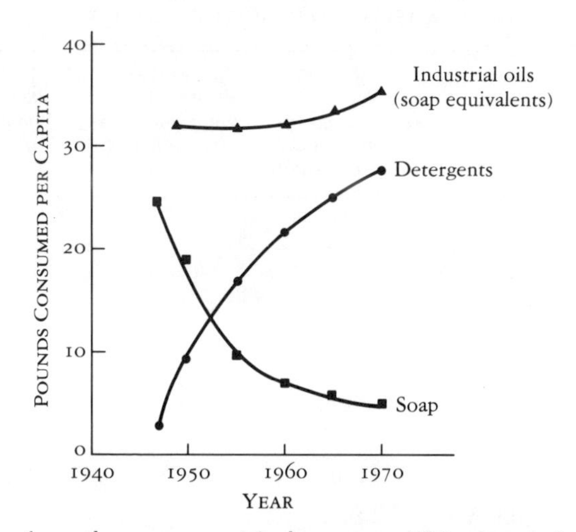

the natural product, soap, with detergent. This shows the relative production of soap and detergents in the United States. We are all about as clean as we were, but now we wash with detergents instead of soap—and soap was a natural product. This is simply to illustrate a basic thing that has happened in the United States, the displacement of natural products by synthetic ones.

Now, I want to discuss the efficiency of natural and synthetic products. Let's take leather as an example, since we have statistics for that industry. I am going to talk about the efficiency of using energy, capital, and labor for all industries.

The average in 1971 for energy was $14 of value added—that's the value of the goods minus the cost of raw materials, or the economic gain of the industry—$14 of value added per million BTU's. This is shown in table 3. The leather industry produced $62 of value added per million BTU's; the petrochemical industry (which makes plastics to replace leather) is composed of the two groups at the top of this chart, and yielded between $1.90 and $4.85 per million BTU's. In other words, leather is a more efficient way to convert energy into value added than is plastics. There is a similar relation with capital. Energy is used to run the machines which are bought with capital, so there's a correlation between the inefficiency of using energy and the inefficiency of using capital.

TABLE 3

LABOR, CAPITAL, AND ENERGY PRODUCTIVITIES FOR SELECTED INDUSTRIES (1971)

INDUSTRY	LABOR PRODUCTIVITY Value added ($) per production worker man-hour	CAPITAL PRODUCTIVITY Value added ($) per dollar of fixed assets	ENERGY PRODUCTIVITY Value added ($) per million BTU's
Petroleum and coal products	28.43	.34	1.90
Chemicals ad allied products	27.75	.80	4.85
Stone, clay and glass products	11.59	.82	7.39
Primary metal industries	11.54	.49	4.42
Paper and allied products	11.40	.58	7.64
ALL INDUSTRIES AVG.	12.43	1.13	14.42
Leather and leather products	6.25	3.64	62.04

SOURCE: U.S. Department of Commerce, *The Census of Manufactures* (published every five years) and *The Annual Survey of Manufactures* (published in the years not covered by the *Census*).

You may say now that the industry that produces plastics uses labor. Well, the first column in table 3 explains it. It is very efficient in using labor. The average figure is $12 of value added per man hour for all industries with only $6 for leather. The leather industry is an inefficient way to use labor. The petrochemical industry is up at $28 of value added per man hour. In other words, when switching from leather to plastic, from cotton to synthetic fiber, from wood to plastics, from soap to detergents, we were switching from a technique of meeting our needs efficiently with energy, efficiently with capital, and inefficiently with labor, to just the reverse.

Figure 18 presents a correlation for various industries between energy productivity and capital productivity, which change together. Figure 19 illustrates the transformation of United States industrial technology since 1948. These bars show annual rates of growth. The rate of growth of the gross national product is around four percent a year. At the top, the fastest-growing production sector is

FIGURE 18

ENERGY AND CAPITAL PRODUCTIVITIES OF SELECTED INDUSTRIES

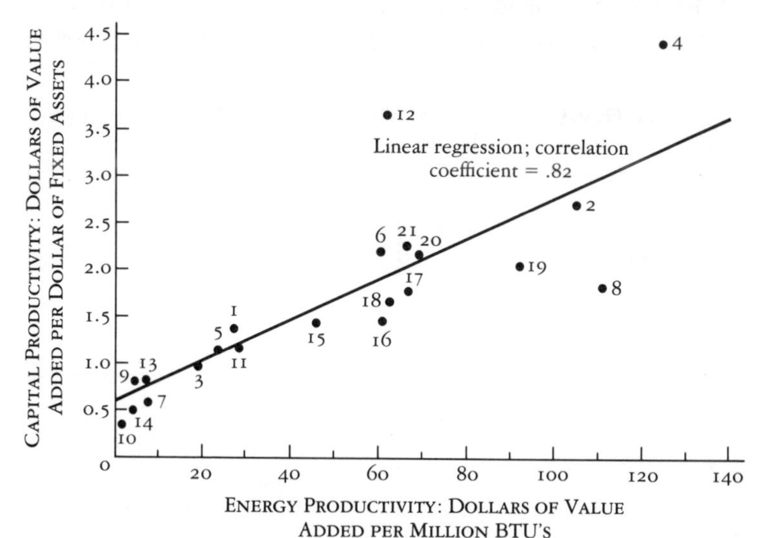

KEY:

1 Food and kindred products
2 Tobacco manufactures
3 Textile mill products
4 Apparel and other textile products
5 Lumber and wood products
6 Furniture and fixtures
7 Paper and allied products
8 Printing and publishing
9 Chemicals and allied products
10 Petroleum and coal products
11 Rubber and plastic products, n.e.c.
12 Leather and leather products
13 Stone, clay, and glass products
14 Primary metal industries
15 Fabricated metal products
16 Machinery, except electrical
17 Electrical equipment and supplies
18 Transportation equipment
19 Instruments and related products
20 Misc. manufacturing industries
21 Ordnance and accessories

SOURCE: U.S. Department of Commerce, *The Census of Manufactures* (published every five years) and *The Annual Survey of Manufactures* (published in the years not covered by the Census). Washington, D.C., U.S. Government Printing Office.

plastic materials, then manmade fibers, and aircraft (the Vietnam War mostly). The biggest single block of industries is the chemical industry. It is the fastest growing. At the bottom of the figure you find industries not growing as fast as the GNP, or that are even shrinking. Lumber, leather, wood, skin leather, wool, and old-fashioned hosiery is at the bottom. What's happened in the United States is that we have displaced those ways of producing goods which are efficient in using energy, efficient in using capital, and inefficient in using labor with the reverse, and the upshot is that we tend to waste energy, to run out of capital, and to run out of jobs. What it does mean to use less labor per unit output, is that there are fewer job opportunities; thus we have created a tendency toward a

FIGURE 19

AVERAGE ANNUAL GROWTH RATES OF FEDERAL RESERVE PRODUCTION INDEXES FOR INDUSTRIES, 1948–1969

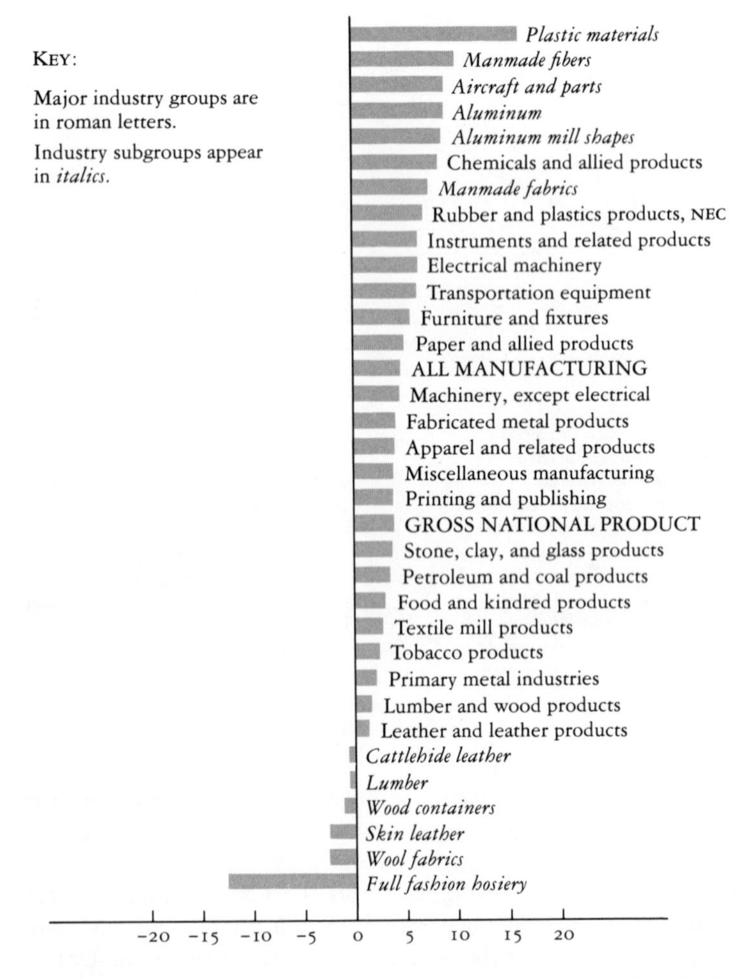

SOURCE: U.S. Bureau of Economic Analysis, *Long-term Economic Growth*, 1860–1970. Washington, D.C.: U.S. Government Printing Office, 1973.

FIGURE 20

UNEMPLOYMENT IN THE UNITED STATES

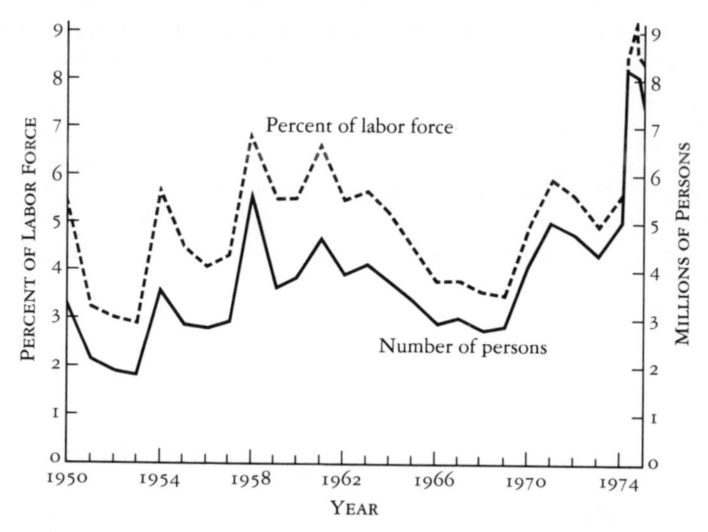

SOURCE: U.S. Department of Commerce, *Statistical Abstract of the United States*, 1975 and 1971. Washington, D.C.: U.S. Government Printing Office, 1971 and 1975.

shortage of energy and a shortage of jobs (called unemployment).

Figure 20 shows recent unemployment figures. When you look at the graph (and you have to remind yourself that this is the wealthiest country in the history of mankind), we have had steady four percent unemployment ever since 1950, and now it's rising. Here I want to quote something. "Unemployment is a ghastly failure of industrial leadership. What is the flaw in the capitalist system which has governed industry for a couple of centuries that it creates and cannot resolve this paradox?" (Cardinal O'Connell of Boston, 1929). That is a very interesting question. In other words, his response to unemployment was to ask why, what is wrong, instead of saying, "Here is my plan," which is what we hear today. I think this is one of the most serious political problems we have. Confronted with a problem like this, people stop thinking; they don't ask questions, they produce answers. I ask, "What is the reason?" Now, I am no economist, and there are various ways of estimating, but I can tell you what some of the latest figures are. Half of the entry of "unemployed" is due to new people looking for jobs, that is, people becoming of age and women entering the job market. And half of it is due to displacement of jobs by new technology. At least half of this problem, therefore, is due to the fact that we have decided to have detergents instead of soap, plastics instead of wood, and you know better than I do that these changes have also caused a great deal of pollution. It's something we have to think about very fundamentally.

Well, there's another side of it. There's a shortage of capital; it's expected that about a third of the expected demand for capital by industry will not be met. I am now quoting reports from the United States Chamber of Commerce and the Chase Manhattan Bank. Mr. Simon, current Secretary of the United States Treasury, has made speech after speech about this. He always has a pat point he makes that in the United States only 17 percent of our GNP goes into

FIGURE 21

RELATIONSHIP OF GROSS FIXED CAPITAL FORMATION AND MILITARY EXPENDITURE TO GROSS NATIONAL PRODUCT

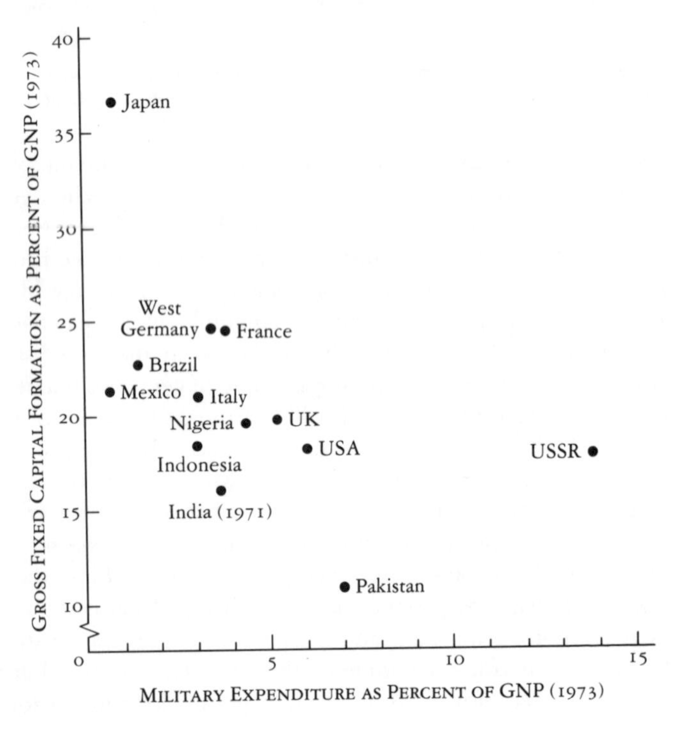

capital whereas in Japan it is 35 percent, and in West Germany it is 25 percent. I point out that he has left something out, namely, that Germany and Japan were forbidden by treaty to have large military establishments; we were spending ten percent of the GNP for military, and they were spending almost none at all.

People seem to misread things I say, and the other day I got a letter from a Danish reader of my book *Poverty of Power* who said I was wrong about that and he had gone and collected data from the United Nations on the relative expenditures for military and capital goods of different countries. He sent me a graph which I reproduce here as figure 21. The graph proves my point. This is a plot of the percent of GNP going into capital and the percent going into military. By and large I would call this a moderately good inverse relationship between the use of capital for military purposes and the availability of capital for industrial investment. It seems to me that when the government talks about capital shortage and the need to cut taxes for corporations, it ought to make the connection, as I have, between unemployment and energy wastage. It ought to make the connection between the shortage of capital and the fact that we have a 110-billion-dollar-a-year military budget. Again, this is something which you learn by looking for these connections.

One of the consequences of all this is that the rate of profit in United States industry is falling. Why? Because the denominator is the assets (the capital) and since we are using more and more capital to produce the same amount of goods, the rate of profit is simply falling. This is illustrated in figure 22. In our capitalist system the rate of profit is the motivation for investment. As a result, there is a lag in investments. Yesterday's *New York Times* (October 21, 1976), again, reported that the recovery in the United States is lagging chiefly because of a hold-back in capital investment. I don't blame the industries because the profits are falling, and the profits are falling because they are using more and more capital and less and less labor. It's a bad situation. Finally, what can you do about it? Figure 23 is a complex diagram, and I want you to look at only part of it. In the center of the figure is the production system that produces value added and it goes to wages and profits. The profits go to taxes and savings. The wages go to consumption and savings, and capital has to be accumulated from savings, either more savings in the bank for someone to borrow or corporation savings which, when it accumulates, becomes capital which is invested in the pro-

FIGURE 22

RATES OF RETURN ON NET CAPITAL STOCK OF NONFINANCIAL CORPORATIONS

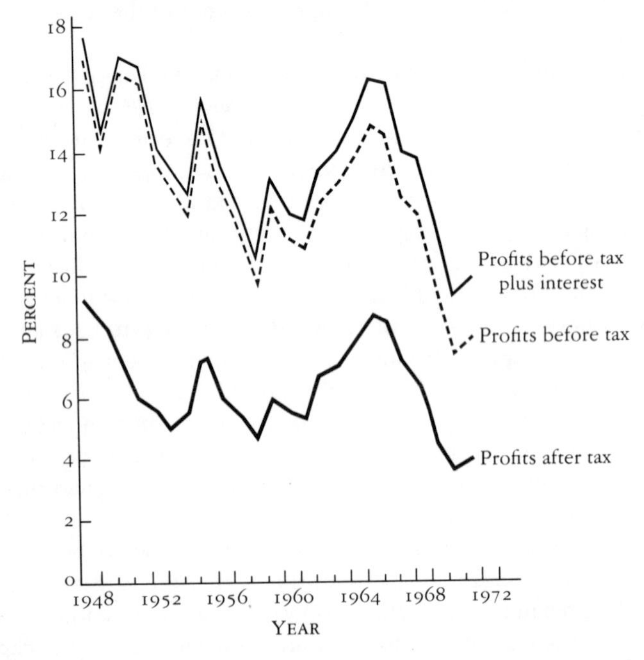

SOURCE: J. A. Gorman, "Nonfinancial Corporations: New Measures of Output and Input," *Survey of Current Business*, March 1972.

duction system. Some of it is invested into making energy which then goes to production. What I have been pointing out to you is that this process of capital yielding value added decreases when you shift as we have from natural to synthetic materials. I have shown you that the efficiency of the investment of capital into energy production falls. What we have is a drain on capital because of the inefficiency of these processes. Big business knows this, and what are they calling for? One of the things they are calling for is a cut in the corporate taxes. That will help—it will mean more profits, more savings, more capital—absolutely! Do you know what else they are calling for? They understand figure 23. They are calling for a cut in consumption. You may not believe that, but it's covered in my book, entitled *Poverty of Power*, and it's really weird. Every *report* on

FIGURE 23

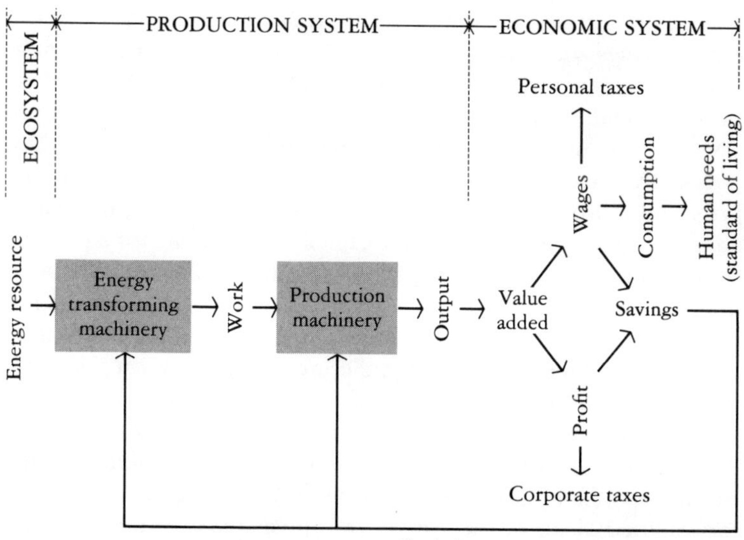

the capital shortage says we must cut consumption because consumption diverts wages from savings, and savings makes capital.

The energy industry is affected worse by the capital shortage than anybody else, and I have explained why. A report appeared in the *Weekly Energy Report* (December 15, 1975), which is the energy industry's sort of Kiplinger Newsletter. The report stated that:

> "Problems with capital formation are tied to social goals: We have been living dangerously for many years by diverting investment funds to meet the operating costs of social objectives," according to Joe Grief, Vice President of Finance for Exxon Nuclear. Less and less of the national output is devoted to increasing the production base, more and more devoted to what Grief calls "social engineering and social enrichment." The result is an economic and social climate "in which the accumulation of savings and therefore the formation of investment capital is discouraged." Says Grief, "Put simply, things have come to the point where somebody must lose. It's a trade-off between more attempts to redistribute income or accelerate savings."

Let me remind you what "redistribute income" means. The conventional idea is to redistribute income from the rich to the poor. That's what Grief means by income redistribution. He said that it's a trade-off between more attempts to redistribute income or accelerate savings, but "you can't do both," said Arnold Pearlman, a young economist with Chase Manhattan Bank. "Income redistribution," he said, "is consumption-oriented. It takes from those who save (and invest) and gives to those who only spend," the poor. Do you know what he's saying? That we should give less money to the poor and more to the rich because the poor don't save, they only consume. Saving is necessary for capital. He is calling for a reverse redistribution. I want to emphasize one thing. This is absolutely logical; there isn't a flaw in that argument, except for a slight problem, since if people don't consume, then who's going to buy all the stuff? Meanwhile, it's a way of gathering up capital. What I want to emphasize here is that we are dealing with a consequence of this process. Ecology is linked up to economics by changes in the production system, which is the link between the ecosystem and the economic system. Capital shortage has brought us to a point where the vaunted claim of the American capitalist system that it could raise the standard of living and make a good profit at the same time has now been given up. It now says that you have to consume less. Incidentally, that is why my most recent book carried the title *The Poverty of Power.* Those who are in power have confessed to the poverty of their power.

If we are going to recognize that it's the character of the production system which governs this relation between the ecosystem and the rest of our lives, *then the final and most basic question is, Who is in control of the production system?* I like another ancient quotation, this time from John Dewey, written in 1953. The reason why I am giving it is that it contrasts so sharply with the pap that we get from our political leaders today. Here's what John Dewey said. "First, the power issue [electric power] is the single most wasteful issue in the political field. The people will rule when they have power, and they will have power in the degree that they own and control the land, the banks, the producing and distributing agencies of the nation." "Ravings," he said, "about Bolshevism, Communism, Socialism are irrelevant to the axiomatic truth of this statement." What he was saying was that there is a question of who is to determine what we are going to do with our production system, and I think that the question exists in a very sharp way today. It is a question which challenges the fundamental precept of the capitalist system, namely, that ayone who owns capital ought to be free to invest it in whatever way gives him the best profit. It raises the issue of whether, *for the sake of the environment and a stable economy, we have to give the people control over production.*

Discussion

GLORIA MACGREGOR: I have pieces of a question to ask. You make the point that producers of natural products such as cotton, wool and rubber are being forced out of the market by low-cost plastic products. Isn't the low cost of such products of social value to the person who could not otherwise afford the product? Mechanical tools such as washing machines, autos, and so on, have increased the longevity of man, it seems to me, as have many drugs in the petrochemical system. Would it be better to have a shorter life span? Kind of tied in with that I would like to ask what you're wearing, because the evidence that we cannot depend on natural fibers is that they have become so high priced that they are not affordable and the customer with more disposable income may actually prefer the synthetic materials since they are so much easier to maintain? How would you care to comment on those?

BARRY COMMONER: You raised a very important point, which is the cost of synthetics. A very strange thing has happened, it's true. When synthetics came on the market they undersold cotton and wool and forced cotton out of production (huge acreage in the United States was cut back—some of it in the South went to soybeans and some of it went to brush), and cotton production actually fell off. When the United States dumped synthetic rubber on the market after the Korean War the Malaysians cut down huge tracts of natural rubber trees because the price fell. In soaps, the whole idea was—and I well remember because we did a very careful analysis of the switch from soap to detergents—that they would get away from high and variable price in fat, which goes through commodity ups and downs. That's why they were interested in switching over to petroleum to make detergents. What has been happening in the last few years is that they have been caught in the rising price of petroleum, and what you find is that the price of synthetics is now close to the price of cotton. Now, in finished goods that may not be true, but mill prices are getting very close. Someone in the business told me the other day that plastic handbags are about the same price of a leather handbag.

You see, you are right. In a sense there is a temporary social advantage in bringing in these cheaper things to people, but the fulfilling of that need was through a system which was bound to go up in price because it was using a non-renewable resource. In other words, if we were to become "addicted" to synthetics, we would be forced from now on into higher and higher prices, and that's exactly what is happening in food. So, temporarily, you are right.

In the area of labor-saving machines, I should have made that clear. No one in his right mind is proposing that the labor-saving machine that uses energy in mining and lumbering and other backbreaking activities is a bad thing. What I am talking about, really, is the difference between, in an extreme example, using a plastic extruding machine to make shoe horns as against having craftsmen carving them out of wood. I see something interesting and rewarding and humane in being a carver of shoe horns. I think it exceeds the humane value of tending a plastic machine that extrudes millions of them and gives off noxious vapors at the same time. So, it seems to me that in certain things, and many young people are moving in that direction, the switch to the machine is not of any social advantage. But in any case, don't you think that society ought to have something to say about it rather than just the man who owns the capital to buy the machine having a voice?

GLORIA MACGREGOR: Well, it seems to me that society has said something about it in the sense that they are buying the products that are produced. We have a lot of examples in the world that show that our so-called emerging nations, which are particularly labor-intensive, have a very low standard of living. One of the things that is very important about the United States is that we have a relatively high standard of living which has been caused in the main by industrialization.

BARRY COMMONER: I am not in any sense talking about turning backward. All I am saying is that when a productive system is designed, it ought to meet certain requirements: compatibility with the ecosystem, efficiency in using non-renewable sources of energy, efficient use of capital, because capital takes away from consumption. And I think that it ought to have the social requirement of providing a certain number of interesting, rewarding jobs for people. One reason why I suggested that is that this is the only way I know to fight the bureaucracy that I mentioned earlier.

Unemployment is a marvelous example. If somebody is unemployed and is going to starve, our morality says that we will not let him starve, and so we set up a social intervention to avoid starvation, and it is called the "unemployment system." What it requires is that a person go to an office, fill out a form—punch cards are made, computers whirl and a lot of clerks operate—and he finally gets a check and he eats. What are we accomplishing? That is an elaborate bureaucratic, expensive way to accomplish the task of providing somebody with food. Now, I suggest a much easier, neater, cheaper, less bureaucratic way, and that is to see to it that there's a job. The bureaucracy involved in finding a job is not that great. If, for example, we had the requirement that new productive instruments allowed openings for people who are ready to enter the productive system, I think it would make an enormous difference. The thing that worries me about unemployment is young people, because at the present time unemployment among new workers, young workers in their teens, is not seven or eight percent, it's around twenty percent, and among black young workers it is forty percent. Now, to me that is a disastrous social situation. Any of us who have had kids grown up to the point where they are ready to enter society—the productive system—the most traumatic moment in their life is when they are ready to test their own capabilities against society's standards, namely, "Can I get a decent job? Will anybody think that I am worth hiring?" I remember my daughter. To have new members of society coming up to the productive system and the productive system saying to forty percent of them, "Sorry, there's no job," that's an enormous social and human cost. I think that it is absolutely essential that there be social governance of the productive system and to see to it that there are jobs for people who are ready to work.

All I am saying is that we have allowed the process to go on because we are unwilling to intervene at the point where the decision is made. We wait until the decision is made in the wrong direction, then we have to set up unemployment, welfare, and various things in order to take care of the people whose hopes are broken by the failure at the point of decision. The EPA wouldn't be necessary if we could intervene at the point where the cars are made and the chemical plants are built, so we wouldn't need unemployment bureaucracy if we could intervene at the point where the production decisions are made.

GLORIA MACGREGOR: How do you account for the fact that not only in unemployment, the problems you are mentioning, or EPA as another good example, it would seem that people are not willing to pay the price of personal interaction, such as I am not willing to pay my neighbor the money to make him ride on the bus to work instead of taking his car. I'm not willing to pay my neighbor enough to shut up his barking dog. I'm not willing to pay him enough to quit using natural resources or other resources, so I say to the government, "You take care of it," because it's kind of costly for me to do it, so we've really increased our bureaucracy.

BARRY COMMONER: I really don't understand that question. I don't understand what you mean by paying a neighbor to shut up his barking dog.

GLORIA MACGREGOR: Well, we're not willing to do that; we say let government do it, and the price of bureaucracy seems to be the price that people want to pay to solve all the problems that come with increasing numbers of people living in a diminishing place.

BARRY COMMONER: Well, that's a very interesting dilemma. We have a free enterprise system, you know. That means that if you own capital you don't want anybody to interfere with what you want to do. For all these other things, however, you want the government to do it for you. I think that's a very unfair division. I think that if we think the government is so good at doing things, why doesn't the government run business too? The other way is that I think this business of the government is not the only way to have social intervention. For example, which government? In my way of thinking, municipal ownership of a power plant is what I would call social control of the means of production, and it does not have to be done by the federal government at all. If, for example, San Jose or Davis wanted to build a power plant—let's say based on solar energy—or wanted to foster the use of solar energy by providing some subsidies or loans, you could do it. That would be social intervention, and you wouldn't need federal government say-so.

I think one of the important things we have to think about is decentralizing public control. For example, one of the most important things about solar energy is that it lends itself to decentralization. There is very little economy of scale in solar devices, whereas, with coal there is enormous economy of scale. It's cheaper to have a big coal plant than a little one. That means that you could compete in the solar industry if you were a small firm. If you owned a supermarket you could afford to build your own power plant to run the supermarket based on solar energy.

We have to get away from the idea that the major agency has got to be the federal government. The other thing I'll say is I don't think we've invented the proper ways of social intervention. All I am saying is we have to confront that problem.

RANDALL FIELDS: This system sure isn't working—for sure—and I understand that, and I have been on very much the same side of these issues for a period of time. It's just that, perhaps, I think there's a different solution, because before I suggest social ownership and social intervention, I want to check with those countries that already have that. For example, the Soviet Union has made some extraordinarily good decisions in the area of pollution control. Recently, the five year plan, for example, adopted by the Soviet Union indicates that in the next five years the Soviet consumer will have to take a little bit less so that heavy Soviet industry can get a little bit more. That's precisely the same trade-off that goes on here. The difference between profit in a capitalist system and the surplus of a socialist system is only a matter of who owns it. There still is a need for investment versus consumption, so that is sort of economic sophistry. It strikes me, incidentally, that even in a democratic, social society like Sweden, remember there was recently a government change over the issue of nuclear power, and guess what they decided once they were in power? Let's go ahead with the nukes. I don't think social control really determines precisely what happens, and I think that clearly the twenty-five years of socialism hasn't been a hell of a lot better than the last twenty-five years of capitalism. They both have a bad record.

BARRY COMMONER: Let me say that in my experience, one of the most insidious political ideas is that the United States needs a model for its political future. I totally abhor the notion that we have to deal with this enormously difficult, complex and unique predicament that we're in by looking around and choosing, "Oh, we'll do that." That's nonsense. What are we here for if not to create our own future?

Having said that, let me say something about the Soviet Union and Sweden. I put it to you very simply. First, Sweden is not a socialist society. In a socialist society, the means of production are owned and governed by the people. In Sweden, certain aspects, the least profitable sector of the production, are owned by society. The rest are given subsidies by the government. The government buys a non-controlling interest. In other words, it is a form of capitalism in which the capitalist is supported by the state, and, in fact, the social security system in Sweden collects money from the workers and turns it over in the form of capital subsidies to the capitalist.

I have described this to the former prime minister of Sweden, Mr. Palme, in just those terms, and he agrees. It is not a socialist society. Yes, there is a socialization, for example, of the railroads in every country in Europe, but the profitable sectors of industry are owned by capital.

I happen to know something about the nuclear power thing in Sweden—I was there just recently. The fact that Mr. Palme and the Social Democrats lost power for the first time in forty-four years over the nuclear power issue had nothing to do with socialism. It was simply that a woman named Brigetta Hambreas, who happened to be a member of the Center Party and a conservationist, decided to introduce a moratorium resolution two-and-a-half years ago in Parliament. Everybody thought it was rather funny, but it sat there and as the campaign developed, the head of her party, Valdine, noticed that every time anybody asked him about where he stood on nuclear power, they were really very enthusiastic. He decided that it would be a good thing for him to be for the moratorium, and he grabbed the issue and ran with it. The result was they won the election. Now you will say, they immediately, then, gave up. Well,

I want to remind you about something. They have a very neatly balanced system there. When the Social Democrats were in power, they were in power only because, I think, of six votes that the Communists gave them. In this election, Social Democrats were for nuclear power and the Communists were against, so even if the coalition won, there was going to be real trouble. Valdine in the Center Party, for his part, was against nuclear power, but his coalition members, the Conservatives and one other party, were for nuclear power. So when he got in office he began to compromise. I just got a letter from a friend of mine which described the whole thing. What he did was agree with the Conservatives. He had to get the Conservatives to back him. He said, "All right, we will fuel the new nuclear power plant that we are about to fuel, but we will not do anything else." He appointed as Minister of Energy an antinuclear man. It was a nuclear fight. Meanwhile, my friend wrote and said, you know, the thing to do is to write to Palme, the Social Democrat, now, because I think that he is going to come out against nuclear power to get back in office.

So, I don't call this a victory for or against socialism; this is Swedish politics.

As for the Soviet Union, it is socialist, absolutely. However, I think it is entirely possible to be socialist and stupid at the same time. I will give you a very simple example. I was in Poland three years ago meeting with the chemical industry of Poland and all the eastern countries. They had just done the dumbest thing that I have ever heard of. There was a material called "Corfam," produced by DuPont, a plastic to be used in shoes. The trouble was that it pinched. The shoes were just no good. DuPont simply closed the factory, so it just sat there. Poland, which has more horses than people, bought the Corfam factory.

Now is that a consequence of socialism? No. I think that it is just a consequence of being stupid. It was, in fact, even a violation of Marxism, because one of the ideas of Marxism is that technology reflects the culture. Now, if there is any technology that reflects the capitalist culture, it is the petrochemical industry. I can point out products which exist, not because anybody wanted them but because they elevated the profit of something else. It's a very fascinating thing, and anybody with a little bit of Marx in his blood would know that, but they just went ahead and did it.

What I want to say is this. If we regard socialism as public intervention in the means of production, I regard that as a necessary but not sufficient condition for having a reasonable system. In other words, it is possible to do un-ecological things if you have a governmental system which doesn't respond well. I know a little bit about what's happening in the Soviet Union. When the whole ecology movement began here, the establishment in the Soviet Union denounced it as a bourgeois affectation. Only in capitalism could something as stupid as this happen. Whoever heard that nature has something to say about what's happening? We dominate nature. That was the line. Along about five years ago, the Italian Communist Party sent a delegation to the Soviet Union to have a discussion about this issue because the Italians thought the Russians were wrong. They got beaten; they couldn't convince them. Two years later they had another discussion, and they split the Russians. Two camps developed: those that said to work with the ecosystem and those who said, "No, we'll dominate." In other words, among two groups in the Soviet Union and among their comrades in Italy, there were different opinions on this issue. All it says is that it is insufficient merely to agree to have a socialist economic system; you also have to understand the production system and the ecosystem. The level of understanding varies.

As far as I am concerned, it doesn't cut any ice to point to the Soviet Union, Cuba, Angola, or China, since they are all doing different things, and say, "See, if you have public control of means of production, it always makes a mess." That may well be true, but it doesn't have to happen the next time, and I think that is what the challenge is. Of course, much more serious is the fact that in the Soviet Union you have the kind of political dictatorship that is abhorrent to us and which violates all the traditions of political freedom we have. Does that mean you cannot have public social control of the means of production without dictatorship? No, it doesn't. Let me remind you that in November, the two largest Communist parties in western Europe, the Italian and the French parties, officially gave up the dictatorship of the proletariat as principle. They said, "No, we will not do it." Now, you may not believe them, but what I am saying is that there is room for difference of opinion here, and I think it's a terrible mistake to regard socialism as being equivalent to what the Russians are doing, what the Chinese are doing, and what the Angolans are doing or what is being done in Cuba.

When I wrote *The Poverty of Power,* the editor of the *New Yorker* was a little leery about my using the word "socialism." I decided it was a scientific word, and the only honest thing to do was to use it and say what I thought it meant. I think it is time for us to begin to think about these things in an open way; it doesn't hurt anybody—no thunderbolts have come down. You see, I think one of the big political troubles in the country, now—this is what you might call the age of "obfuscation"—is that people don't want to talk about these simple ideas. Many places I go people say, "Oh, you don't think that the United States is a capitalist country?" That's utter nonsense; of course it is. They say, "Well, you see we have a post office." Then someone says that we should look at how badly the post office runs. Well, you know, if the post office weren't carrying all the junk mail the capitalists put into it, it might run a little better!

FRANCIS DUBOIS: I think that there are some facilitators in the audience, and I would like to ask for a question or two from them to Dr. Commoner if he is willing to answer them.

AUDIENCE PARTICIPANT: Would you say that there is a point of no return where the production system loses its ability to change itself?

BARRY COMMONER: Well, I wouldn't put the question quite that way. I'd ask the question, "Is there a point where the faults in the production system will paralyze its ability to produce?" I think there is, and I think we came close to it in 1929 and 1930 when the production system, because of a failure in the economic system, slowed down enormously. I think this is a problem which we are beginning to face. In yesterday's *New York Times* (October 21, 1976), there was a long article about whether the decline in the economy is a long-term decline. Various economists were quoted as suggesting that maybe it is a long-term decline, that there is something fundamentally wrong in the system which is beginning to paralyze its ability to grow. The growth is essential in order to make up for the shrinkage of jobs and the shrinkage of the efficiency of capital, because the only way you can keep people at work is to make the thing grow. But you are making it grow by introducing exactly the capital goods that shrink employment and waste capital, so it is an endless fight.

I think, myself, that there may well be an inherent instability in the system, but you have to ask the question: Why has the system done so well? I think that there are some good reasons. One is that an awful lot of the strain was taken off by allowing the production system to chew up the ecosystem. It represents a kind of buffer. Now, one of the reasons why we are in trouble, is that the environmental movement has closed the door. There was a very interesting case in Pittsburgh where there was a big fight about coke ovens. They cause cancer and pollute the environment, and there's no way of putting any environmental controls on them. They have to be completely rebuilt. I believe that it was U.S. Steel that just agreed to an EPA court action to completely rebuild all their coke ovens in Pittsburgh over the next ten years. It is going to cost six hundred million dollars. They issued a statement which said that they would do it, it would cost six hundred million dollars, and it would prevent them from expanding their facilities, because that six hundred million dollars would have gone into further expansion. In other words, that investment reduced the productivity of the capital. What they were saying was: We will take six hundred million dollars and, instead of investing it in a way that would produce more steel, we're going to invest it in a way that will clean up the environment. But we don't make money from cleaning up the environment; we make money from steel, and that's going to hurt, to clean up the environment. You have to recognize that insisting on capital investments to control pollution means a lowering of the productivity of capital, which just worsens the problem. So, we are closing the door on the present system and changing the status quo.

The other thing that is happening is that the unions are beginning to insist on environmental control in the work place. For example, when you start in the chemical industry and begin to insist that all hazardous operations be closed down or cleaned up, then they are closed.

Now let me give you another simple example. Refineries used to be shut down once a year to be cleaned. Now they are so expensive that it costs too much money to shut them down except every two or three years. The result is that repairs are made while the refinery is operating, and the union, the oil, chemical, and atomic workers, now, are complaining bitterly that they are exposed to much more hazardous conditions because they have to work at it while there is a valve leaking and something may explode and so on. They are insisting that they don't want to operate under those conditions. That's putting a constraint on the efficiency of that capital.

Or, take the business in Lordstown, Ohio, which is the most advanced automated plant for the auto industry. When it was opened, one-third of the cars that came off were sabotaged. That is literally true. The workers sabotaged them. They could not stand the pace of the assembly line, and they had to change it. That means that there is a limit to the increased labor productivity in the auto industry. So, there are now new constraints which I think may lead to an economic collapse, and we have to start thinking of other ways of doing it. The reason why I like to bring up the alternative is so we aren't like "rats in a trap" when all of this falls on our heads. Is there no other way to do this? Well, we ought to think about it.

FRANCIS DUBOIS: I find that we are getting down to details with this discussion, and I would like to entertain just one more pertinent/impertinent question, if you will.

AUDIENCE PARTICIPANT: Would you make some comments on the use of energy in Sweden, please. I believe they waste less energy.

BARRY COMMONER: Oh, sure, we have elaborate ways of wasting energy! For example, we don't have mass transit. Every European country has a much, much better railroad system than we do. If you take Italy, which everyone says is on its last legs, you can take a train between Rome and Florence which is as good or better than our best Amtrak, and, as you are riding it, you will see a half-mile away a brand new concrete track being laid down for a super 115-mile-an-hour train to go between Rome and Florence. Are we doing that? Electrified trains, you know, are the most efficient way to travel. So, one of the reasons for energy efficiency in every country in Europe is transportation. Another is insulation. They have very heavy insulation and very rigorous regulations requiring it, so that Swedish homes have always been very heavily insulated. Another thing is the distances. You see, we have a large country. Instead of doing the sensible thing of breaking the country up into sort of autonomous regions, where, for example, people drink beer from a brewery which is close enough so you can afford to return the bottles and wash them, we don't do that. There used to be nine hundred breweries in the United States, and now there are about twenty. That's why we have throw-away bottles, because you can't afford to ship the empties back. In other words, the centralization of industry in the United States has led to an enormous increase in shipment, and that is a big waste of energy. When I was in Harvard, I well remember that the hills in western Massachusetts all had pastures and dairies, and the dairy industry on the East Coast has now essentially disappeared. The milk comes from Wisconsin—burning gasoline all the way. That sort of thing is much more limited in Sweden and most European countries, and I think that is where an awful lot of the saving takes place there.

II

INTERMEDIATE TECHNOLOGY AND THE INDIVIDUAL

E. F. Schumacher
Sim Van der Ryn

INTRODUCTION: *Yvonne L. Hunter*

Introduction to E. F. Schumacher

Yvonne L. Hunter

E. F. Schumacher has been described as the John Maynard Keynes of the post-industrial society, the spirit of Thomas Jefferson or Ben Franklin, the guru of intermediate technology and deserving of the Nobel Prize.* For the man who has spearheaded what may be one of the most profoundly important movements of our time, these descriptions are apt. The author of *Small Is Beautiful: Economics As If People Mattered,* Schumacher's main thesis is that economics, in developed as well as developing nations, should be concerned with people, not goods. He argues that in many situations, high (advanced) technology is not appropriate and may even be harmful to the people and the environment. An economic system that makes work meaningless is not doing people much good, even if it feeds them. Finally, Schumacher is convinced that the world must get back in touch with its more spiritual needs. All of these beliefs lead to an incredible array of associated or spin-off ideas, activities, projects, and related philosophies—in short, the intermediate technology movement.

Schumacher's background is one that lends credibiity to his ideas. He is no mere armchair critic of contemporary society, but a trained economist with a variety of practical experiences. Born in Germany in 1911, he was a Rhodes scholar in economics. During World War II, along with many other Germans living in England, he was interned and then released to do farm work. As he describes it, the farm work brought him a renewed awareness of the importance of the connection of people and the land, a connection he feels is

*Peter Barnes, "Wise Economics," *New Republic,* June 15, 1974, pp. 29–31.

almost lost on the millions of city dwellers and urban intellectuals. After the war, he became an economic advisor to the British Control Commission in postwar Germany, and from 1950 to 1970 was the top economist and head of planning at the British Coal Board, one of Europe's largest industrial enterprises. Against this orthodox and seemingly conventional training and experience, is another side of Schumacher. He was president of the Soil Association, one of Britain's oldest organic farming organizations; founder and chairman of the London-based Intermediate Technology Development Group; a close student of Gandhi; and a sponsor of the Fourth World Movement, a British-based campaign for political decentralization and regionalism. In the mid-1950's he was an advisor on rural development to the prime ministers of Burma and India.

In 1966, Dr. Schumacher founded the Intermediate Technology Development Group, an organization that specializes in tailoring tools, small-scale machines, and methods of production to the particular needs of a nation or region. In 1973, *Small Is Beautiful* was first published; it was released in paperback in 1975. A sleeper publication at first, it is now widely read in the United States and abroad. It has been called an "intellectual catalyst" and its effects have been profound. Consider California governor Jerry Brown's statement at a press conference in 1976: "If you want to understand my philosophy, read this," he said, pointing to a copy of *Small Is Beautiful.* The United States Congress has recently taken up the ideas of intermediate technology, based in large part on Schumacher's philosophy, by writing encouragement and dollar amounts into the budgets of various agencies, such as the Agency for International Development, the Community Services Administration, the National Science Foundation, and the Energy Research and Development Adminisration.

Some critics present substantial arguments against Schumacher's ideas; others nit-pick, often misunderstanding the thrust of his thesis. An example of such criticism is that described by Philip M. Boffey, writing about the 1977 Annual Meetings of the American Association for the Advancement of Science.* In addition to the considerable enthusiasm at the meetings for Schumacher and intermediate technology, Boffey recounts an argument presented that

*Philip M. Boffey, "AAAS Meeting: Drought Was the Topic of the Week," *Science,* vol. 195, no. 4282, p. 966, March 11, 1977.

intermediate technology is an "'intellectually empty' haven for disenchanted middle-class youth who are seeking a 'playground' for their hobbies and a new 'radical chic' cause now that the Vietnam War is over." In addition, Boffey states that the critics not only contend that the various solar houses or composting toilets built by intermediate technologists are actually "unattractive," they assert that the American public will never prefer such products to those produced by our large-scale, centralized technical system. While it may be correct that some (but certainly not all) intermediate technology–type products lack aesthetics, the remainder of the criticism seems a bit harsh. It is the sort of conventional wisdom that Schumacher and others in the movement are trying to bring into question.

In order to understand Schumacher fully, one must understand the full meaning of intermediate or appropriate technology. It has been called gentle, soft, non-violent, ecological technology. Sim Van der Ryn, State Architect and head of California's Office of Appropriate Technology, opened the conference with an informal discussion of the meaning of intermediate technology. The remarks included after Schumacher's question-and-answer session are offered to give the reader a background for the material to come. They are based on testimony Mr. Van der Ryn gave to the State Subcommittee on Land Use Planning in November 1976.

Like the panels organized for the conference, Schumacher's ideas can be broken down into a number of broad categories: appropriate technology, ethics, agriculture, third world development, scale of technology, and energy. Using the concepts presented in *Small Is Beautiful,* the panelists provide their own insights and thus present an in-depth discussion of Schumacher's basic ideas.

E. F. Schumacher is a visionary. His theories are considered both unorthodox and all-encompassing, weaving together threads from Galbraith and Gandhi, capitalism and Buddhism, science and metaphysics. Perhaps, as Harold Gilliam suggests in an article that appeared in the *San Francisco Chronicle* after the conference, Schumacher is somehow not as close to John Maynard Keynes and the great economists as he is to Benjamin Franklin and Thomas Jefferson.* They too were visionaries.

---

*Harold Gilliam, "Technology's Middle Zone," *San Francisco Sunday Examiner & Chronicle,* March 6, 1977.

Intermediate Technology and the Individual

E. F. Schumacher

Tonight I wish to talk about civil and human liberties. And I want to do so in relation to a subject that has been very prominent in this conference, namely, technology.

This word "technology" simply means know-how, knowledge of how to do things. It is not confined to hardware. Intermediate technology, appropriate, soft, adaptive, progressive technology, whatever you call it, can apply to education, to medicine, to government, to industry, to anything you like.

I shall take as my starting point, as it were, a remark made by one of the most influential thinkers of the nineteenth century, most influential whether we like it or not, a certain Dr. Karl Marx, who said something to the effect that freedom is the recognition of necessity. Freedom is the recognition of necessity, or you might say, freedom springs from the recognition of necessity. We humans live under many necessities which appear to deny our freedom, but we can, nevertheless, be free provided we have the insight to recognize the necessities: and then can act freely. Some people I know interpret this Marxian saying as a counsel of despair, as an invitation to accept everything in a spirit of fatalism. But if Marx was a fatalist, we are all fatalists. This is nonsense. No, the recognition of necessity enables us to act freely and effectively; to disregard necessities means to act ineffectively, and breeds nothing but frustration and impotent rage.

What can we do to maintain and to increase the range of personal freedom? In England, it used to be said that people were free to dine at the Ritz or to sleep under the bridges of the Thames. This is the kind of philosophy that enraged Karl Marx and many others. You don't have to be a Marxist to get angry at that. Now, why are these people who are "free" to sleep under the bridges of the Thames actually forced to do so? They somehow don't fit into everyday present society. This is the point. If more and more people don't fit and are forced to rough it, we have to take thought. After all, we have an immensity of ability, an immensity of resources, and should be able to solve the problem of poverty at the drop of a hat. There are always going to be a number of misfits, but I am not talking about them. I'm talking about millions—fifteen million unemployed, for instance, in the western world, who do not seem to fit into the system. They are free, or forced, to "sleep under the bridges." Even in your great country we are not dealing with misfits, we are dealing with a failure of society. A society that fails in this way is in imminent danger of losing its freedom. And the loss of freedom always stems from a failure to recognize necessity.

It is now necessary, in my opinion, to recognize a number of things. First, we cannot establish justice and social harmony simply by welfare. I'm not saying anything against welfare, but it isn't enough. Second, the only effective way of helping people is to help them help themselves. Just to do things for them demoralizes them. Thirdly, you cannot help people to help themselves if all you can offer is advice on what they could do if they were happy, healthy, highly educated, affluent, rich people. If we are serious about extending civil and human liberties, we must, above all else, understand what it means to be poor, disadvantaged, neglected. Only on a deep and sympathetic understanding of the constraints of poverty can a healthy and just society be built.

All right then, let us try and understand the constraints and conditions of poverty. I will not do the work for you. It is of value only if you do it yourself; in your imagination, try and live this situation; try and understand the constraints of poverty. As Gandhi recommended, if you are uncertain about these things, remember the poorest and most miserable person you've met in your life, only excluding mental defectives, and then say, how is what I am now doing relevant to his condition? How is it likely to affect him or her? If you do this, you may come to unsuspected insights and conclusions. For instance, you may find what you are doing is totally irrelevant to that condition. All right. We write him off. Forget about him. But he can't write himself off. Oh, you may say, "What I am doing is something that really emphasizes how far he has been left behind and increases the distance." O.K. The gap widens. Let the devil take the hindmost. But if the hindmost doesn't want to be taken by the devil, we may have to use power to keep him down. Or you may find, "What I am doing is something designed to help him to help himself, so that he may establish himself with self-reliance, self-respect in freedom."

Now I do not wish to suggest that this social, political concern is all that matters. There is also a non-human reality producing necessities which we must recognize if we want freedom. The necessity to maintain or regain ecological balance, the necessity to fit ourselves into a more modest framework of energy usage, the necessity to achieve a more rational pattern of settlement than that which has been emerging in the last fifty years—vast congestion on the one hand and great emptiness on the other—the necessity to revise our methods of production so that people can enjoy their work and are not drained of their humanity by doing it. Mindless work creates mindless, unhappy, and most likely violent, truculent, disruptive people. It is a great danger to any society.

Now you may say, all this is a very tall order. How can we do all these things? Indeed, it is a very tall order. But I am talking about necessities, which if we do not recognize them and act accordingly, will destroy our freedom. I'm not talking about mere desirabilities. In a field as big as this, everybody is invited to make his contribution. My own contribution, as I fancy, is quite a modest one. It's simply this. I say that a great deal of what has occurred has happened because of a specific development of technology. There is the excitement of discovery. There is, of course, the profit motive. But whatever the origin of this development, which is now encompassing the whole world, the fact is that technology has developed in a way which leaves out the majority of the population of the earth. Actually, it has left out the majority of the population in the most successful countries. And, it has missed out on certain options. Some of you have heard me talk about this before. I will repeat it. There is a strange phenomenon which we have to understand: I call it, in my unacademic language, the law of the disappearing middle. You see, to start with, there is a very primitive, stage-one technology, maybe just tools. That's fine. It will always be there; it's unforgettable. Then there come innovations—stage two. Splendid! We have stage one and stage two, side by side. Then there come further innovations. Excellent! Then something strange happens, namely, stage two disappears. That is not very noticeable because one, two, and three are pretty close together. But with rapid technological development, let's say we are now at stage twelve; between one and twelve there is hardly anything left. It's disappeared. Please, don't just listen to what I saying, try to verify it in your own experience. As the chairman mentioned, some thirty-odd years ago, I was a farm laborer. We had excellent equipment. Very cheap, simple equipment. We did good farming in England. All this equipment that I worked with doesn't exist anymore. It has disappeared. And now to farm these three hundred acres you do not have to have $5,000 to invest in equipment, you have to invest $100,000 or maybe even $200,000.

If you go into a book shop, you can buy the latest and the best publications or the classics. The "middle," anything published in say, 1965, is unobtainable. The same with technology. And this produces the tremendous strain, in my opinion, in this modern world, not only in the western society, but also throughout the Third World, where the poor are fixed on stage-one technology, which isn't good enough; they can't make stage twelve, and the middle is empty.

We might say that the number of options has diminished. It's either later or nothing. If you've made it to stage twelve you are all right. If you haven't made it, you are all wrong, and you are frozen in that situation. And this applies not only internationally, it applies also in our own countries. The gene pool, as it were, is not now rich enough to solve our problems. We know how to do things, big, complex, with a lot of capital and a lot of violence. But when it comes to doing things in outlying areas, in rural places, with people without a lot of capital and sophistication, we don't know how to do it. *We don't know.* For many decades we've been offering the Third World technologies which only the rich among them can use. Hence, we get the nasty remark that foreign aid is a process of collecting money from the poor people in the rich countries to give it to the rich people in the poor countries. Not because we are people of ill will, but because the kind of thing we are able to offer can be received only by people already rich or powerful. We don't know—except some of us do—how to do things so they fit within the boundaries of poverty, whether the poverty is in our own country or in other countries. We don't know how to do things with a minimal ecological impact, and with a minimal use of scarce fossil fuels. All that we don't know. But that is what the majority of mankind needs to know. And the majority of mankind do not have the intellectual and other resources necessary to create this knowledge. If we know that we don't know these things, then we know what to do. If we don't know that we don't know these things, then we are just tossing about causing confusion. As soon as we, the rich, realize that we don't know how the poor can help themselves, we can learn something about it. The great achievement of western civilization is not landing people on the moon, a somewhat pointless activity, but the discovery of a methodology of learning and of problem solving. The knowledge that now is required is how people survive in conditions of poverty and how they can get out of them.

I think that concern for the tensions in our society and concern for the poor is the primary obligation that educated people ought to accept. When I was working for the Coal Board, I recognized that the intelligence of modern society goes preponderantly, almost exclusively, in the direction of solving the problems of the rich, but not of solving the problems of the poor. I recognized that the "middle" was lacking, that for the poor there was nothing but the most simple technology, which was not enough for a good life.

So, I set up an organization, without money, called the Intermediate Technology Development Group, Ltd. Still very limited. And I am happy to say that an intermediate technology organization has now also been set up at Menlo Park in this country. We are discovering that many people, many groups of people, are already at work. Once we wake up to the real problem of poverty, which cannot be solved by transferring the technology of the rich and cannot be solved simply by welfare, and once we realize that many people are already concerned with this, then the task is to do a certain job of networking. This is, in fact, what the Intermediate Technology organization at Menlo Park is doing with consummate skill. It is very humble work, but it is relevant to the maintenance of freedom in society, not the freedom to sleep under the bridges of the Thames, but a freedom of self-reliance, of self-respect.

Ladies and gentlemen, I am not in the entertainment business. I'm always on the commercial line. If I am standing here talking to you, it's not just to spend an evening with you. It is a plea to help us, or to help Menlo Park; to help the people who are trying to face the future. Now I am quite certain, although certainty in this world is a relative term, that the future will not be the same as the past. I am quite certain that after many decades of work on the subject, that the fuel crisis is not just a little incident. It is *from now on.* I don't know how much time we still have. That's an idle question. I know that we have to find alternatives to the ways we are doing things now. We have to get busy now. Not to say, "Oh well, this will come in the next century and we can live it up, the same way we have done it until October 1973."

I would just like to make this remark. I've noticed that most of us take our present life for granted. We think this is normal. But the moment we look back into history, and not far back, we realize how abnormal it is. We realize how we are taking for granted things that have only come yesterday. I'll give you one or two examples. We get very upset when suddenly we are confronted with a situation of energy shortage. The OPEC's General Secretary has been traveling around ever since the late 1960's telling the oil importing countries, "For goodness sakes, mitigate your requirements. At the rate you are going, we are emptied out and in twenty or thirty years it will be finished. What happens to us, us the oil exporting countries? Back to sand and camels." Nobody listened. We take things for granted which are the most abnormal things in history. Fifty years ago, and there are many people I can see here who can remember fifty years ago as I can remember it, the use of oil in the world was five percent of what it is now. Five percent! Most of it was used in the United States. If you take all ocean-going transport now, and put oil on one side and everything else on the other, in terms of ton-miles the oil is twice as much as everything else put together. Fifty years ago there was none of that. The total ocean-going transport of oil then was just about the same amount as oil spillage now.

It is by taking cognizance of facts such as these that we may come to realize how abnormal our life has become and how uncertain that it could continue. I'll give you one more fact. If we think we can resume the path that we followed before October 1973 and within the next fifteen years still want to have about twenty years of proved oil reserves so that we can do some planning, then we'd have to discover two new Alaskas every year! Now you tell me what the chance of that is. There is no chance of it. In other words, what I am saying is quite simply this. That society won't come to an end. Life will continue. But the exaggerations, the absurdities, that we have allowed to grow up in the last thirty years, they can kill us.

And so we had better take thought of the morrow and adjust ourselves to a different world. In a sense, the party is over. Now, those who have realized this, have tried to get down to work; which means getting down to some reality, not simply espousing ideas, prophesies, and forebodings. Ideas have to be incarnated in real things and the incarnation has to have something to do with the way people make a living. And this means technology.

Therefore, all over the world, organizations are springing up now that call themselves appropriate technology—or intermediate technology or soft technology or energy-saving technology or ecological technology—groups. The name doesn't matter. These are attempts to reorient our thinking; not to think simply that the future will be the same as the past, only more so. For this work, we need help. Many people are doing it on their own, in their own backyard. But I don't think there's enough time to do it in the backyard; I think we have to do it with the establishment. That's where the knowledge is, where the economic power is.

I lay myself open to political criticism when I say this. I know that. They say, "Oh, they won't let you." Well, I think all of this is a mythology. The establishment actually consists of people like those here in this hall—people who are not blind. They see the problems as we are seeing them. They may be working in such a framework that they have little elbow room, having to abide by all kinds of rules. But you can always find thoughtful people in all walks of life. The message for the future, I'm quite certain, is this: Don't talk to universities, don't talk to government, don't talk to corporations. Talk to people! They may be connected with government, with universities, with corporations—so much the better. They are approachable, and they are just as worried as we are. People say: The multinational corporations, which recent technological developments have created, are the enemy. Or government is the enemy. I suggest to you, ladies and gentlemen, we can't afford now to have any enemies. We have to get to work and to work with all people of good note.

This work is not actually very expensive. I would not recommend to switch all research and development expenditure to this new work, but say five to ten percent. And then it turns out—and now I'm not talking merely theoretically, but from practical experience when I say—it can all be done. Modern knowledge is extremely flexible, extremely adaptable. If we decide to create a technology which will really see us through all these difficulties, it can be done. Let's try it. Let's go ahead with it. Let's get out of the situation which has been described as using all our intelligence for rearranging the deck chairs on the *Titanic.*

This is a sales talk because I do believe time is short. And I am not, as I said, in the entertainment business. I make an appeal. Join us in this work. Thank you very much for your attention.

Discussion

AUDIENCE PARTICIPANT: You have said, and I believe in the conference today the primary emphasis was, that an intermediate technology was needed primarily for the poor who could not afford the more expensive technologies. And I wonder if you would agree that for moral and conservation ethics whether it is also necessary for those of us that can afford it that we move down to an intermediate technology?

E. F. SCHUMACHER: Yes, exactly. In fact, forgive my saying this, but this is precisely what I just said. I said there are these sociological things that force us to rethink, but there are also necessities we have to recognize coming from nature. So I entirely agree; you are exactly on the right line.

AUDIENCE PARTICIPANT: About education. You wrote in your book that it's the most valuable resource in your view. And what I got from that is what we need to do in education is reorient from a number of fixed ideas that were set out in the nineteenth century and to begin to develop human potential in thinking clearly, thinking in terms of self-reliance. In America we have a decline in the quality of education especially on the secondary levels. A decline in the scholastic aptitude scores for instance. The taxpayers of Oregon are in a revolt, so-called, over their lack of basic education in schools. How do you feel your theories fit into the educational field? What about the quality of education?

E. F. SCHUMACHER: We have in Britain, and I also understand here, gone for big, big schools, losing sight of human proportions, and it's been terrible. But the content of the education is more interesting. I don't know about you, but I myself hated school. And when I left school I said I would never fall into this old-fogey saying, "Well, school was the happiest time of my life." It was the worst time of my life. I thought it was phoney. I don't know what to do about education, but I can look around and see elsewhere what is being done about education. Now, for instance, in China, they worked out that it costs thirty peasants working for a year to keep a young person at a university. And when that young person has been at university for five years, he or she has consumed 150 peasant work years. And then, an old brute by the name of Chairman Mao said, "What do the peasants get back from it?" He said, "You'd better go and serve the peasants in some remote village." These young educated people then came into the village and they found they were completely useless. The uneducated local midwife could help a woman give birth to a child while this educated person couldn't do it because he can only do things when he is surrounded by $20,000 worth of equipment. And then these young people came back and said, "Look, you are making fools of us." And so this combination of work and learning has had a magical effect on the curriculum of Chinese education. Well, similar attempts of reorientation have been made in certain African countries, like Tanzania, and so on, how to get education back to real life. And I think we'll have to search. Schools ought to be more than a place to keep the children out of my hair.

AUDIENCE PARTICIPANT: I wonder if you would talk about the Tropics, where the great majority of the underdeveloped countries, and I presume of poor people, are living today, particularly with respect to agriculture.

E. F. SCHUMACHER: Intermediate technology or any one phrase like this cannot describe the totality of it. What is it? In my approach, it is simply to say, let's ask which technology, what type of technology fits the situation? You know, there I have a great advantage because fate had given me the name of a shoemaker. And a shoemaker isn't a good shoemaker if he just makes good shoes. He has to know a lot about feet. You see, it used to be said that the Soviet Union had produced the best boots ever and set up a most marvelous factory and they made 500 million pairs of these boots. All the same size! They threw them at the population and said, "You can like it or lump it." Now most of them had to lump it because the shoe didn't fit. So now where we really have to have compassion and patience and understanding with what we are trying to do, is to say what fits now in these precise conditions. This cannot be just a standardized job. We may know what may fit into the conditions of Chicago or Los Angeles, but we don't know how to make things that fit into conditions in Bismarck, North Dakota, let alone what fits the conditions of Tanzania or Zambia or what-have-you. There is no glib answer to be given; we just have to ask ourselves what is the requirement in this precise situation. And there we do very well to listen very carefully to what the local people are saying. Once we have listened to the people, we can take it home and use our knowledge, which is superior to the knowledge normally available in developing countries.

AUDIENCE PARTICIPANT: Dr. Schumacher, it seems to me that in order to recognize these necessities we talked about tonight, in our behavior and throughout our society, we have to address a question dealing with virtuous behavior. How can a society come to express this virtuous behavior? Can this kind of behavior be learned, must we learn it by following somebody else's example? Does the government have a role here?

E. F. SCHUMACHER: You cannot get virtue just by preaching. Maybe you can get it by example. The nearest thing I can come to it for what you call "virtuous behavior" is if people get into touch with reality, that may save them. The human being is a being of brains and hands. The great fault of modern education is that it only appeals to the brain, the memory and such things. But what it actually means to grow things, to be in touch with nature, this is very rarely taught. Maybe it is better in this country; it is certainly very bad in my country. So maybe the only way to regain some virtue is to combine learning and doing, learning and working, so that by education we don't become haughty.

AUDIENCE PARTICIPANT: Dr. Schumacher, you have stated that we are in a stage-twelve technology and that we started out in a stage-one technology. I'm not really certain what the stages are in between. Could you give some concrete examples of the middle that has disappeared?

E. F. SCHUMACHER: Let's take a situation in which people need materials for building houses or other things. They have in East Africa, Southeast Asia, everywhere you can see it, extremely primitive arrangements to fire bricks. It is really very primitive. It uses a lot of fuel and produces very bad bricks. They know this isn't good enough, so they go to the World Bank or some other aid-giver and say, "We need a brick factory." What they are given is a vast brick factory, producing two million bricks a week, all in one place. It's wonderful and uses fifty jobs. So, they build this factory and then they find the results. Good Lord, two million bricks a week in the same place! You just can't use them. You have to have lorries and roads to carry it all across the country. It breaks their back.

Now we at Intermediate Technology take a different view. We say there are these appalling little brick works all over the place. We must go in and make a good job of it with the height of 1977 knowledge. It costs hardly any capital, they don't get into debt, employment is maintained, productivity is increased, it incorporates local production for local use, and no great transport. It is infinitely better than the huge brick works in the outskirts of Teheran or Caracas.

AUDIENCE PARTICIPANT: Would you comment on the concept of the middle axiom?

E. F. SCHUMACHER: In large-scale administration, whether in business or in government, you are in a dilemma. You think something is necessary and ought to happen, so you pass it on. If you pass it on as an order, well then, you lose the active and the enthusiastic participation of the lower levels. If you don't do it, you are blamed for not doing your job. Well, what do you do? What do you do? You have to hit upon what I call the middle axiom. Namely, to elucidate information which makes people see what they ought to be doing. This is always a work of art. It is always, as most of life is, a squaring of the circle, where if you have to do the impossible thing you have to have total freedom. And you have to have control. This is what it refers to. Now here again, it's always a red-letter day when you can hit on some idea. An example I give is statistics. Most statistics are worthless. But you can make the lower levels in an organization connect certain statistics not because I want to know it, but because they ought to see it. And when they see it, then they will act without my orders. As some person has said, and I hope I won't get into trouble with IBM, the most intelligent use of the computer is to order a computer, then to say the computer is coming, and let's rethink our organization and exactly what happens. And when you have done all the programming work, cancel the order. That is what I mean by the middle axiom.

AUDIENCE PARTICIPANT: Would you comment on the statement on page 57 of your book concerning the role of women, the need for outside jobs, and how this relates to a Buddhist philosophy?

E. F. SCHUMACHER: Your question is quite justified, and since I've had this question about 175,000 times in the United States, and no other place, all I can say is that after this experience I would have formulated the statement more carefully. I was reporting, of course, what would arise out of a Buddhist philosophy. However, I still think that mothers of young children are better in looking after the young children than anybody else. I think it is a bad use of resources for these mothers to go into factories, to produce normally useless things, while the children run wild. Now if a lady comes to me and says, "But I hate children, I'm sorry I ever became a mother, I don't want to look after them, I can make arrangements that they will be looked after by somebody else, and I want to go into a factory," I would say, well go. In the book, I was just explaining where the addiction to modern growth economics is—the production seems to be the thing, and not the education of children. This is wrong, and the Buddhists wouldn't have it. I'm all in favor of people getting what they need. And I think this one remark should not delay us too much, because after all, we are not here to discuss certain preoccupations of feminism. But I am just a humble European. I don't encounter this at home. I'm with women who know their part with mankind. They don't have to have it affirmed in words, because we are able to affirm it in deeds.

AUDIENCE PARTICIPANT: What can we do for ourselves, what guide can we use?

E. F. SCHUMACHER: The first thing I can see, the first thing is that every one of us has to clear his own mind. We have to go into ourselves and say, "Do I believe all of this? Do I think, for instance, that something really has to change, or am I one of those who believes it is all a lot of nonsense? Do I think it is more important to have a society, to have harmony on this earth, to live with a degree of equality and rights, than to have a man on the moon? What is our sense of value?"

To sort out all of this is the first stage for every one of us. The second stage is, if we believe that there is some truth, we have to find some reorientation. Well, who is doing anything about it? Join them. Join them. Countless people come to me and say, "We need a new type of agriculture, but I am too small, I can't do anything about it." I say, "Are you a member of the Soil Association?" I'm talking about Britain now. They say, "No." "It only costs you five pounds." "Oh, well, um." They don't even do that which they can do immediately. So the second thing is, after you have cleared your mind, join those people who are already doing something.

The third thing is, when you have joined, work. Whatever you can do, it may be very little, but everybody can do something. And if you don't find anything that is worth joining, initiate something. It's not very difficult to initiate anything. All you need is a brass plate and a letterhead and you are in business. It isn't very much, but it's a darn sight more than just coming to conferences and talking about it. So it isn't very difficult. Let's take a leaf out of Gandhi's book, "A step at a time is good enough for me." We are not sent to save the world, we are sent to do the right thing. But to do the right thing we have to use the magical word, Stop! That's all.

What Is Appropriate Technology?

Sim Van der Ryn

Appropriate technology, what is it? There are a lot of different words and terms that we've heard over the years. Some call it gentle technology, some call it soft technology, or intermediate technology.

The idea of appropriate technology is based on a view of the world that says industrialized societies ought to be using their remaining stock of non-renewable resources to build a society which can sustain itself when remaining stocks of minerals and fossil fuels can no longer be economically extracted and converted to use.

Appropriate technology puts a value on long-term stability based on living patterns and technology that harmonize with and maintain the health and viability of life-supporting ecosystem processes. Appropriate technology looks to new possibilities growing out of an era of limits. The potential of appropriate technology is to indicate ways through to new possibilities that are practical, low cost, create new jobs and do not degrade the environment. Appropriate technology is the means to make a transition to a high quality of life, a steady-state society that is our only alternative as the fossil-fuel age comes to an end.

There is currently much discussion about what kind of technology is appropriate to a high-quality, ecologically harmonious, humanizing steady-state society. I see three characteristics of any appropriate technology. They are syntropy, coherence, diversity.

Syntropy is the tendency to minimize the amount of waste heat in converting from one form of energy to another or in performing useful work. All forms of energy conversion, be they natural or man-made, are entropic. In other words, energy or matter cannot be transformed from one state to another without resulting in a great deal of waste heat. Natural ecosystem processes are inherently syntropic and efficient, that is, they capture solar energy and convert it to other usable forms of energy, such as plant or animal biomass, through complex chains of biophysical events. Man has yet to improve on nature, although the availability of plentiful high-grade fossil fuels has blinded us to the entropy law. Non-syntropic technology based on extraction and degradation of the environment is thermodynamically inefficient in balancing potential energy to useful work. Appropriate technology matches the properties of energy sources with end uses to achieve desired results with least waste. It is thermodynamically absurd to set off a controlled nuclear reaction to superheat steam to 1200° to generate electricity to heat water to heat the air in this room to 72° to keep the surface of our skin at a comfortable temperature. The practical, appropriate, and syntropic way to heat space is with the sun, since solar energy can be directly transformed into usable room temperature heat. And, to build so that the need for heating and cooling is minimized. Some have called the present times an Age of Affluence; more accurately, it might be dubbed an Age of Entropy.

The concepts of syntropy and entropy are significant in choosing strategies of energy conservation. It takes energy to provide energy in a usable form. As our readily accessible supplies are exhausted, more and more energy is required to extract, convert, transport, and store energy. For some proposed technologies such as oil-shale extraction, more energy is used to extract a unit of usable fuel than is produced. In assessing energy sources and conversion technology, the measure of value is net energy, the quantity actually available for use after paying all the energy costs to bring it on the line.

A second criterion is coherence, wholeness. What this means is that a particular technology is in tune with and can coexist with the syntropic behavior of natural systems. Only in this way can a high quality of life be sustained. When technology disrupts and distorts the chain of ecosystem events, the result is an accelerated cycle of more disruption, more technology, more costs. Again, our technology is still immature and poorly integrated with the complex biology of living processes. Appropriate technology does not look backward; it seeks a higher, more mature and scientific integration of human needs with ecological necessity. In a coherent system there is no waste, since waste is only a resource out of place, and each byproduct feeds into another system. Some examples of an inappropriate technology are sewage treatment, air conditioning, and air pollution.

The third criterion for an appropriate technology is diversity. What diversity gives us is the possibility to adapt to localized and changing conditions. We have seen that ruthless simplification and uniformity in a technology imposed for short-run advantage, an organizational convenience, has disastrous effects as conditions change. Nowhere is this more true than in agriculture. Some examples are the Green Revolution and the Sahel.

Again, the principles of diversity follow from an observation of efficient natural process. A diverse, decentralized technology is adaptable to change and is more likely to be syntropic. An example: last night as I wrote these words in my office, 20,000 square feet of space were brilliantly lighted, although I was the only one in the office. Why? To save in building cost there is only one light switch for the entire floor.

A few more comments on what appropriate technology is and is not. The analogy for appropriate technology is the dynamic biological system, as opposed to the stationary mechanical system that still controls so much of our technological and organizational thinking. We tend to see appropriate technology as small rather than big; as elegantly simple as opposed to dumbly complex; as crudely right as opposed to precisely wrong; decentralized rather than centralized; low cost rather than expensive; and employing capital and fossil fuel sparingly. Higher efficiency means more modest capital and energy demands to create each new job. However, the appropriate technology concept is evolving, and we need to avoid dogmatism. The concept of what is an appropriately small scale depends on context. For example, while it might be technologically possible to make every home self-sufficient electrically through renewable energy transformers such as solar cells or windmills, this would be absurd. The criteria are syntropy, coherence, diversity. The scale of any life-support technology will depend on what is necessary to maintain a stable, healthy community and ecosystem structure.

Appropriate technology does not mean "low technology" as opposed to "high technology." Much modern technology, such as that used in communications and electronics, lends itself to small-scale applications, and enhances the prospects for diversification and decentralization. However, communications and electronics are certainly not examples of low technology. The implementation of appropriate technology is a way of providing non-inflationary growth in a diversified responsive society. Using appropriate technology provides for sustained yield from our natural resources and gives people greater individual control and responsibility, a buffer against disruptive and uncontrollable results of fossil fuel shortages, and the skyrocketing costs of immature, highly entropic technologies.

III

AGRICULTURE AND INTERMEDIATE TECHNOLOGY: LABOR, CAPITAL, AND ENERGY-INTENSIVE METHODS OF PRODUCTION

Peter Gillingham
E. Phillip LeVeen
Vashek Cervinka
Michael Perelman
Roger Garrett

INTRODUCTION: *Yvonne L. Hunter*
MODERATOR: *Richard Rominger*

Introduction

Yvonne L. Hunter

As moderator Richard Rominger has said, there are two ways of looking at the relationship between intermediate technology and agriculture. One is from the perspective of developing countries, "where they do not have a technological agriculture built up as we have in developed countries." The other is from the perspective of developed nations, "where we have a high degree of technological agriculture, but where we are now running up against the constraints of diminishing resources and ecological balance."

Schumacher contends that we must insist on returning to the land the nutrients we remove from it when we harvest its crops. This leads to a much different view of agriculture, not only in developing nations, but in developed nations as well. Incorporating Schumacher's overall theories, it implies increased organic inputs, labor intensiveness, and small, decentralized systems of agriculture.

How, where, and when these alternatives are applicable forms the basis of much of the debate between the proponents and critics of different agricultural systems. Briefly, the critics of modern technological agriculture maintain that it is inappropriate for developing nations, and, because of natural limits, may no longer be appropriate in developed nations. On the other side, critics of alternative agriculture movements maintain that the present high technology methods are necessary to sustain productivity, feed the world's hungry, and provide the diversity of foods to which we are accustomed.

More specifically, critics of the Schumacher approach fear that such solutions will not work, that they will not provide the continued abundance and variety of food available in developed nations, and they will not feed the world's hungry. Writing in *The Nation,* Asher Brynes states that "we cannot wisely consider things that we don't have enough time to do." Today, he says, we only have time to feed the world, not to experiment with ways of improving nutrition, agricultural methods, and one's spiritual outlook. Brynes concludes that like Norman Borlaug's Green Revolution, which is based on high-energy, industrialized agricultural systems, Schumacher's intermediate technology approach to agriculture simply will not pan out through the difficult times we face. Unfortunately, Brynes does not suggest any alternatives other than what seems to be just muddling through.*

This view, like others, possibly looks to Schumacher and intermediate technology to be all things to all people in all situations. What most critics forget is that Schumacher is not advocating a return to primitive agriculture, but is advocating implementing our highest scientific and technological skills to develop technology that is appropriate to the situation. Peter Gillingham, Schumacher's close associate, speaks to this issue in his presentation. He says,

> The relevance of intermediate technology to this discussion lies in the belief that by scaling down from overdevelopment and scaling up from underdevelopment to appropriate and efficient levels of technical sophistication we can begin to find more sustainable, environmentally and humanly sound ways in which to carry on the functions necessary for our survival.

It is necessary to get away from extremes in arguing the merits of intermediate technology, agriculture, and our current crises. It is necessary to critique the current situation and to suggest alternatives, directions, and compromises. For the most part, the speakers on the agriculture panel do just that.

Peter Gillingham begins his presentation first with a discussion of the problems of modern agriculture and continues with a discussion of some alternatives.

Philip LeVeen takes an analytical approach in critiquing alternative agriculture, emphasizing that Schumacher "gives too little attention to structural and institutional factors in his analysis of the reasons for our present predicament." LeVeen goes on to examine and describe three major structural impediments to the adoption of

*Asher Brynes, "Yes—But What Do We Do, Now, about Famine?" *The Nation,* June 8, 1974, pp. 725–726.

intermediate technology: competition in family farming, the incentives to use labor-saving technologies, and how, in the context of California agriculture, organized labor and farmers both have a stake in the continued substitution of capital for labor.

Vashek Cervinka examines problems of production, productivity, labor, energy, and farm technology in developing countries. Cervinka critically examines different agricultural models and concludes that an adaptation of the model of European agriculture from 1910 to 1940 may be a practical model for agriculture in developing nations.

Michael Perelman takes modern, large-scale, high-technology agriculture to task, stating that large farms are the most "God-awful agricultural system ever devised by the human people." Perelman's is an extreme view, but one that accurately capsulizes many of the criticisms of modern, high-technology, high-energy agriculture.

Finally, Roger Garrett discusses "appropriate engineering" and its role as producer of solutions for specific problems. He states that what may appear to be inappropriate technology or poor engineering today may be simply obsolete technology—a result of changing conditions. Garrett argues that good engineering will continue to produce appropriate technology for varying conditions in time and space.

The panelists provide a critical look at all types of agriculture. Their thoughts should be valuable to those who are considering the alternatives, those who value the status quo, and those who simply are seeking to expand their understanding of the complexity of our agricultural dilemma.

Appropriate Agriculture

Peter Gillingham

Agriculture in today's world is an enterprise of many faces. From the slash-and-burn techniques practiced through much of the equatorial belt to the terraced rice fields of Japan, from the agrarian communalism of China to the modern agribusiness of the American Midwest, there exists a wide range of methods by which man provides for his daily sustenance. In this country the dominant form of agricultural practice is typified by a capital, energy, and resource intensiveness unrivaled in technological sophistication by any other food-and-fiber-producing system in the history of the world. The relevance of intermediate technology to this discussion lies in the belief that by scaling down from overdevelopment and scaling up from underdevelopment to appropriate and efficient levels of technical sophistication we can begin to find more sustainable, environmentally and humanly sound ways in which to carry on the functions necessary for our survival.

The modern American agricultural system has weaknesses deriving in part from its own real successes, but alternatives already exist which can strengthen or compensate for these weaknesses. It is to these alternatives that this paper addresses itself.

Some potential problems in the path of American agriculture may be quickly enumerated. Much of the production capability of our farming system rests in dependence on petrochemical-based fertilizers, pesticides, and herbicides.[1] Since these materials' derivation depends largely on fossil fuels, there is ample reason to be concerned for the sustainability of this system of agriculture (natural

---

This article was prepared by Michael Shepard and John Jeavons. Numbered references will be found at the end, beginning on page 98.

gas supplies pose a particularly acute problem since they are due to run out in about a decade).[2] By the year 2000, twenty-five percent of all the energy consumed in the world in 1973 will be required just to produce nitrogen fertilizer.[3] Soil depletion in the form of microbial life, nitrate and salt accumulation, and loss of organic matter and plant nutrients are other factors which often accompany present practices.[4]

We should all be aware that the now largely barren and denuded lands of Iran, Iraq, Syria, Lebanon, Palestine, Algeria, Tunisia, and other countries once supported thriving civilizations which drew their strength from the health and richness of their soils. Through neglect of the earth they tilled, the base of their cultures crumbled.[5] The United States, revealing dangerously similar neglect, has lost at least one-third of its topsoil since the Revolutionary War, and is at present losing five billion tons of topsoil a year, or an average of twelve tons per acre of agricultural cropland, from wind and water erosion. Most of this could be prevented by modified agricultural practices. While American agriculture proudly boasts of its high yields, it is becoming increasingly costly (in terms of dollars, resources, energy, fertilizer inputs, and environmental impact) to raise the yield enjoyed by our farmers. Decreasing marginal returns on chemical inputs into crop production indicate that petrochemical agriculture is approaching the upper limit of its effective yield potential. [6] Yet we find that for virtually every crop grown in this country there are farmers in some other nations who, on the average, are getting yields that are twice as large.[7] The tendency of other countries to adopt agricultural practices similar to our own is not always well advised. If the world population were to be fed a high protein and calorie diet such as ours, and were to use agricultural technology employing sunlight and petroleum as the only sources of energy for food production, the usable known reserves of petroleum would last only thirteen years.[8]

Fortunately, solutions to America's agricultural problems exist. We can produce as much or more food and fiber of at least equivalent quality while using less energy, money, and capital, and we can produce our food in a way that enhances rather than degrades the environment.[9] That these solutions would make food production more sustainable in the long term should be of interest to agricultural researchers, farmers, and all of us who eat food. Basic to the solution is the premise that the health and fertility of the soil, its living community and its natural processes, must be enhanced. Modern agriculture, in contrast, often regards the soil simply as a medium to hold the crops while they absorb chemicals, water, and sunlight.

A detailed description of the operating or proposed alternative systems that might be regarded as compatible with the intermediate technology approach is beyond the scope of this paper. Those systems rely primarily on productive capacities of people and nature, avoid techniques that deplete or paralyze either, and are more suitable to the modern world than primitive technologies, while conserving more resource and capital than most modern technologies designed according to earlier unconscious assumptions of limitless factor availabilities. A few examples will perhaps serve to indicate why these systems may deserve wider interest and more intensive exploration than they are now receiving. Organic tractor farming is a system fairly close (at least in outward appearance) to current large-scale farming using chemical inputs. The primary difference is that no fertilizers, pesticides, or herbicides are used. Natural fertilizers such as livestock manure and green manure wastes (abundant in the United States) are substituted for chemical nitrogen, thus enhancing soil structure and organic matter content. Crop rotation and interseeding of leguminous crops such as winter vetch serve to maintain nutrient levels, prevent erosion, and encourage microbial life. A recent study by William Lockerets, Barry Commoner, and others compared organic with conventional farms in the corn belt and found virtually no difference in economic returns per acre from crop production.[10] Organic farmers often enjoy yields as high or higher than their conventional counterparts. Organic farms on the average use one-third as much energy—other than ambient sunlight—as conventional farms. This is due primarily to the large amounts of energy needed to produce the chemicals used by the conventional farms. Since operating costs are a smaller fraction of the total value of production for organic farms, the Lockerets study concluded that organic farming would be less vulnerable to declines in crop prices such as the one that occurred between the 1974 and 1975 seasons.

The French Intensive Biodynamic Method is a technique of farming drawn largely from practices used by the Chinese for forty centuries. Contemporary research indicates that this method should produce, on the average, two to six times the United States national average of protein-source beans, grains, and rice, and two to sixteen times the vegetable and soft fruit yields. It appears that these yields can be accomplished while consuming as little as one-half to one-sixteenth the water and nitrogen fertilizer used in traditional methods, and one one-hundredth the human and mechanical energy used in traditional methods, once the soil is in balance. Preliminary research indicates that using the French Intensive Biodynamic Method one farmer may eventually be able to earn $8,000 a year working twenty hours a week on a one-tenth acre plot. A full balanced diet may eventually be grown on as little as 2500 square feet per person in a six-month growing season. Five to ten years of research probably remain to realize the full potential of this method.[11]

Since the French Intensive Biodynamic and other similar biointensive methods are not at all suited to the highly mechanized techniques of most food producing systems in this country, the objection is often raised that they can do little more than increase productivity of small gardens, while contributing little to the overall food production of the nation. In fact, a large proportion of the arable land in this country today is in areas poorly suited to large-scale mechanized agriculture. One-quarter to one-third of our potential cropland, or one hundred million acres, is found in urban fringe areas where development pressures are rapidly threatening its value as food-producing land.[12] Of the 740 million acres currently in use as pasture and rangeland, a substantial proportion will be used in a more cost-effective manner through intensive farming. The entire nation could be fed by small farms on the close fringes of urban areas if six to twenty-four percent of the population were to use biointensive techniques within small radii around major population concentrations in ways that could provide livelihoods for millions. This would also change the trend toward profit-driven land speculation that has been separating the cities more and more from food-producing land.

The relationship between agriculture and intermediate technology is inextricably connected with the trend, relatively unquestioned during the past half-century until the recent past, toward removing the small and family farmer from the land in favor of much larger, highly mechanized operations. There are many who feel that we should now attempt to reverse this trend—to encourage and support the family farm and small farmers' cooperative marketing organizations as a viable and integral functioning unit of our society. This is the very heart of the Jeffersonian vision of free rural land holders, and was expressed nearly a century later by another American statesman, Abraham Lincoln: *"Population must increase rapidly, more rapidly indeed than in former times, and ere long the most valuable of all arts will be the art of deriving subsistence from the smallest area of soil.* No community whose every member possesses this art can ever be the victim of oppression in any of its forms, such Community will alike be independent of crowned kings, money kings, and land kings."[13] Yet another voice, that of the Congress of the United States, was added to the list of supporters of the small landholder in the Agriculture Act of 1961: "It is hereby declared to be the policy of Congress to . . . recognize the importance of the family farm as an efficient unit of production and as an economic base for towns and cities in rural areas to encourage, promote, and strengthen this form of farm enterprise."[14] Heeding Lincoln's characterization of small farming as an art and the Congress's recognition of its vital importance to the nation's well-being, we must begin to support and encourage the small farm not merely with rhetoric but with action as more than an economic unit. The small farm represents a way of life that must be preserved.

The small farm, all factors considered, is actually a more efficient unit of production than its large-scale counterpart. To quote Sterling Wortman of the Rockefeller Foundation, "Most large-scale mechanized agriculture is less productive per unit area than small-scale farming can be. The farmer on a small holding can engage in intensive high-yield 'gardening' systems such as intercropping, multiple cropping, relay planting or other techniques that require attention to individual plants. The point is that mechanized agriculture is very productive in terms of output per man-year, but it is not as productive per unit of land as the highly intensive systems are."[15] Much of the economic advantage large farms now enjoy is not due to any streamlining inherent in their size, but in the tax structures and subsidies which favor corporate "tax loss" farming, land speculation, and vertical stratification of the food industry from production of fertilizers, to farming, to packaging and distribution of food products. A recent director of agricultural economics for the USDA stated that "we know from our studies in the Department of Agriculture that the rates of foreclosure and delinquency are greater on big farm loans . . . than for smaller loans on family farms." Yet, credit is hard to come by for the small farmer. One study by the legislative reference service of the Library of Congress concluded that farms with over $40,000 annual sales would face proportionately larger financial difficulties if price supports, direct and indirect, were discontinued by the government.[16]

Excessive mechanization and size are not vital to successful farming. Measuring the degree of mechanization by the amount of fuel consumed per farm worker, we find that smaller farms produce crops of equal and often greater dollar value per acre, on land often less fertile than the major large farms, while consuming less energy and requiring less mechanization in the process.[17] In Egypt and Taiwan small farms very often produce higher yields than their large neighbors and large farms in other nations.[18] At least part of the higher efficiency of small farms is due to the increased concern the small farmer has for the land he works.[19] Corporate farmers, in contrast, are sadly out of touch with their land. A vice-president of Tenneco for Agriculture and Land Deveopment has said that at his company "we consider land as an inventory, but we are all growing things on it while we wait for the price appreciation of development. Agriculture pays the taxes plus a little."[20]

A strong argument for decentralization of food production and distribution near market areas arises when we see that "twice as much energy is consumed outside the farms in processing, packaging, transporting, distributing, refrigerating, and cooking food as the farmers consume in growing it. . . . The energy requirement for the entire food system could in the long run, present a more serious problem than meeting the energy needs for crop production."[21]

Disappearance of small farms is a tragic phenomenon for two reasons. Displaced farm workers find it difficult to play productive roles in society; their skills and training are of little worth in the city, where they all too often end up adding to the welfare rolls and tax burdens. Also, the positive human values of the small farms are being lost. A study comparing two farming communities in California's central valley, one dominated by large farms and the other by small family farms, found that where the smaller farms prevailed the community enjoyed a higher standard of living characterized by superior physical facilities such as streets, sidewalks, parks, stores, retail trade, twice the number of organizations for civic improvement and social recreation, and so on.[22]

The facts are these: (1) the smaller farmer can be more viable; (2) other countries have found it possible to get double the typical United States yields; (3) multiple cropping without pesticides reduces pest losses more than pesticides do in monocropping[23] and without damage to desirable life forms;[24] (4) there is enough organic matter produced in this country each year virtually to eliminate the farmer's dependence on fossil-fuel energy and chemical fertilizers;[25] (5) many methods use less water and fertilizer while producing higher yields with greater protein levels than does the present system.[26] All these facts, and more, point to opportunities for a new, higher yielding, resource conserving, sustainable, modern agricultural system. The solutions involved are all ones which we could begin to implement tomorrow if the will of the people and the creative expertise of our agricultural research establishments were channeled into appropriate and constructive action.

REFERENCES

1. Wilson Clark, *Energy for Survival,* New York: Anchor Books, 1974, 652 pp. Clark points out that instead of being an energy-producing system, agriculture consumes the equivalent of 150 gallons per person annually, and is the largest consumer of petroleum in the nation (p. 168).

Barry Commoner, *The Closing Circle,* New York: Bantam Books, 1972, 344 pp. "The use of synthetic organic chemicals and pesticides rose *495%* and *217%* respectively, between 1946 and 1968, while the use of chemically fixed nitrogen fertilizer rose by *534%* in the same period" (p. 172). In Illinois in 1945 ten thousand tons of chemical nitrogen fertilizer were used to produce corn yields averaging 50 bushels per acre; in 1965 *four hundred thousand* tons were used to produce 95 bushels per acre (p. 81). "Every year a bigger and bigger dose of chemicals is necessary to sustain monoculture crop yield—at the expense of the entire system's stability" (p. 174).

Roger Revelle, "The Resources Available for Agriculture," *Scientific American,* September, 1976, pp. 164–178. "It is estimated that by the year 2000, 160 million tons of chemically fixed nitrogen—four times the 1974 figure—will be used in world agriculture, requiring 250–300 million tons of fossil fuels" (p. 168).

David Pimentel, et al., "Food Production and the Energy Crises," *Science,* November 2, 1973, pp. 443–449. Sixty to eighty percent of the corn yield increases since 1940 were due to energy inputs (p. 445).

David Pimentel, et al., "Land Degradation: Effects on Food and Energy Resources," *Science,* October 8, 1976, pp. 149–155. "Despite the fact that from 1949 to 1969 a net 15% (58 million acres) of U.S. cropland was annually withheld from production, total crop output increased 50%;

meanwhile population increased 30%. An increase in productivity of about 6% per year more than compensated for the loss of cropland to highways, urbanization, and other special uses. *Except for the elimination of some less productive lands, all of the factors contributing to increased productivity on slightly less land required significant increases in the use of fossil energy*" (p. 149).

Richard Merrill, ed., *Radical Agriculture,* New York: Harper & Row, 1976, 459 pp. See chart showing increase in chemical inputs, p. 297.

2. Revelle, "Resources Available," p. 168: "The present high prices and foreseeable exhaustion of fossil fuels raise serious question of whether the energy intensive agriculture of the developed countries can be extended to other parts of the world or can be continued for very long in any country."

Pimentel, et al., "Food Production," p. 443: "Fossil fuel inputs have in fact become so integral and indispensable to modern agriculture that the anticipated energy crises will have a significant impact upon food production in all parts of the world which have adopted or are adopting the western system."

John Jeavons, director of agricultural research, Ecology Action of the Midpeninsula, Palo Alto, California: personal communications about natural gas supplies, January 1977.

Pimentel, et al., "Energy and Land Constraints in Food Protein Production," *Science,* November 21, 1975, pp. 754–761. "To gain some idea of what the world energy needs would be for a high protein-calorie diet if U.S. agricultural technology were employed, an estimate is made of how long it would take to deplete the known world reserves of petroleum. The known reserves have been estimated to be 86,912 billion litres. If we assume that 76% of the raw petroleum can be converted into fuel, this would equal a usable reserve of 66,053 billion litres. If petroleum were the only source of energy for food production and if we used all the petroleum reserves solely to feed the world population, the 66,053 billion litre reserve would last a mere 13 years" (p. 758).

3. Amory Bloch Lovins, "Energy in the Real World," *Stockholm Conference ECO,* San Francisco, December 13, 1975, p. 9.

4. Pimentel, et al., "Land Degradation," pp. 150, 152: "During the last 200 years, at least a third of the topsoil on the U.S. croplands has been lost. . . . Gross soil erosion in the U.S. is about 5 billion tons annually. . . . On the basis of our estimate of an annual gross loss of more than 5 billion tons of topsoil . . . estimates of about 12 tons per acre appear to be reasonable. . . . About 200 million acres were ruined or seriously impoverished for crop cultivation by soil erosion before 1940 . . . this nation's land continues to be eroded. . . . Sediment damages (three quarters of which come from agricultural lands) are estimated to cost the U.S. $500 million annually. . . . Soil sediments, the associated nutrients (for example, nitrogen, phosphorus, and potassium) and pesticides have an ecological impact upon

stream fauna and flora. The added nutrients may increase aquatic productivity, resulting in eutrophication; in contrast, when suspended sediments are present they reduce light penetration, which reduces the productivity of aquatic ecosystems. Fish food may then be less abundant. . . . The decline of crop rotation and increase of crops grown in continuous culture (such as corn) for example, have increased soil erosion. . . . Soil erosion loss in western Iowa is up 22% because of current U.S. farm policy to increase food production."

Barry Commoner, "Nature Under Attack," *Columbia Forum,* vol. 2, no. 1, Spring, 1978: "The organic content of our midwest soils has declined in the last 100 years by about 50%."

Wm. A. Albrecht, "The Half-Lives of Our Soils," *Natural Food and Farming,* September, 1976, pp. 7–11. Studies at the Missouri Experiment Station show that the use of chemical fertilizers speeds up the loss of soil fertility (p. 9).

Sulphur Institute Journal, Fall, 1971, p. 16; Michael Blake, *Concentrated Incomplete Fertilizers,* London: Crosby Lockwood, 1967, pp. 14 ff. Large yields induced through heavy fertilizer applications upset the balance of nutrients in the soil and induce deficiences in our foods.

Michael Perelman, "Efficiency and Agriculture: The Economics of Energy," in Merrill, ed., *Radical Agriculture,* pp. 26–27.

Richard Merrill, "Toward a Self-Sustaining Agriculture," in *Radical Agriculture,* p. 299: Runoff from nitrogen fertilizer is very high: in addition to eutrophication problems, nitrogen toxicity is a problem: "(1) Nitrates may react chemically with blood hemoglobin and impair the circulation of oxygen in the blood or cause vitamin deficiencies. . . . (2) Nitrates may react with amines in the body to form nitrosamines and related nitrosamides. These compounds are known to induce cancer. Concern that increased use of nitrogen fertilizer may be linked with the growing incidence of cancer in modern society has been expressed by even the medical establishment." There is also the danger of nitrogen contamination in aquifers: "The rate of water recharge from deep percolation is so slow that the possible nitrate pollution of aquifers . . . will take decades. However, once nitrate gets into the aquifer, decades will be required to replace the water with low nitrate water. . . . By the time the trend was established, a dangerous situation could be in the making that could not be corrected in a time shorter than it took to create."

Sir Albert Howard, *An Agricultural Testament,* Emmaus, Pa.: Rodale Press, 1972. 253 pp. "Salt accumulation is not just a result of too little rainfall, it is the result of low porosity and permeability" (p. 147). "Excessive irrigation without adequate drainage and aeration, overcultivation, use of artificial 'manures' all contribute to alkali accumulation" (p. 152). "Excessive development of alkalis in India as well as in Egypt and California is the result of irrigation practices modern in their origin and modes and instituted by people lacking in the traditions of the ancient irrigators who had worked these same lands thousands of years before. The alkali lands of today, in their intense form, are of modern origin, due to practices which are evidently inadmissible, and which in all probability were known to be so by the people whom our modern civilization has supplanted" (p. 155).

5. Vernon Gill Carter and Tom Dale, *Topsoil and Civilization,* Norman: University of Oklahoma Press, 1976. In Palestine "enough rain falls in most areas to nourish the crops grown by the ancients. As long as the soil was capable of absorbing the rainfall, this was a relatively prosperous and beautiful country. In the 1st century A.D. Josephus described the land of Galilee, Sumeria, and Judea: '... the soil is universally rich... these countries are moist enough for agriculture and very beautiful. They have an abundance of trees, both wild and cultivated, that are full of fruit. The land is not naturally watered by irrigation, but chiefly by rainfall, of which they have no want'" (p. 85). Today the "Promised Land" is a sad commentary on man's stewardship of the earth.

6. Clark, *Energy for Survival.* The efficiency of nitrogen fertilizer input/output crop ratio declined fivefold in the period from 1949 to 1968. "Every year a bigger and bigger dose of chemicals is necessary to sustain monoculture crop yield—at the expense of the entire system's stability" (p. 174).

Pimentel, et al., "Land Degradation," pp. 153–154: "Although agricultural technical changes have more than offset the potential productivity loss due to soil erosion, the costs in terms of reduced potential food productivity and increased use of energy have been high.... Based on the fact that at least a third of the topsoil on the cropland has been lost, and that for each inch of topsoil loss there has been a corresponding decrease in productivity, we estimate that the production potential of U.S. cropland has been reduced ten to fifteen percent.... A total of 2.1 billion gallons of fuel equivalents annually has been used to offset past soil erosion losses in the U.S. This amount of fuel is equivalent to 50 million barrels of oil annually, or about 4% of the nation's total imports during 1970.... To feed a growing U.S. population or increase the world per capita diet (or both) the amount of cropland under cultivation can be increased or the productivity of the land already cultivated can be increased. Either course would require enormous amounts of energy and could not be continued indefinitely.... To feed the world population, projected to increase to 6 to 7 billion in less than 25 years, food production must be about doubled on the available arable land.... To double the world's food production on current land resources would require about a three-fold increase in energy for agriculture within less than 25 years." (This last statement must be assuming that U.S. technology is being employed, since more intensive methods could increase yields while reducing energy inputs.)

7. John Jeavons, personal communication, January, 1977.

8. Pimentel, et al., "Energy and Land Constraints," p. 758.

9. John C. Jeavons, 1972–1975 Research Report Summary on the Biodynamic/French Intensive Method, Palo Alto, California: Ecology Action of the Midpeninsula, 1976, 17 pp.

William Lockerets, Barry Commoner, et al., *A Comparison of Organic and Conventional Farms in the Corn Belt,* St. Louis, Missouri: Center for the Biology of Natural Systems, 1975, p. 27.

Cedar Rapids Gazette, May 23, 1976, p. 21B.

Howard, *Agricultural Testament.* "Japan provides perhaps the best example of control of soil erosion in a country with torrential rains, highly erodible soils, and a topography which renders the retention of the soil on steep slopes very difficult. Here erosion has been effectively held in check by methods adopted regardless of cost, for the reason that the alternative to their execution would be national disaster. The great danger from soil erosion in Japan is the deposition of soil debris from the steep mountain slopes on the rice fields below. . . . For this reason the country has spent as much as ten times the capital value of eroding land on soil conservation work, mainly as an insurance for saving valuable rice lands below" (p. 143). "If we regard erosion as the natural consequence of improper methods of agriculture, and the catchment area of the river as the natural unit for the application of soil conservation methods, the various remedies available fall into their proper place. The upper reaches of each river system must be afforested; cover crops including grass and leys must be used to protect the arable surface whenever possible; the humus content of the soil must be increased and the crumb structure restored so that each field can drink in its own rainfall; over-stocking and over-grazing must be prevented; simple mechanical methods for conserving the soil and regulating the run-off, like terracing, contour cultivation and contour drains, must be utilized. There is, of course, no single anti-erosion device which can be universally adopted. The problem must, in the nature of things, be a local one. Nevertheless, certain guiding principles exist which apply everywhere. First and foremost is the restoration and maintenance of soil fertility, so that each acre of the catchment area can do its duty by absorbing its share of rainfall" (p. 147). Howard cured the problem of alkali at Pusa and in the Quetta Valley through "attention to soil aeration, to the supply of organic matter, and to the use of deep rooting crops like lucerne and pegeon pea, which break up the soil" (p. 153).

10. Personal communication from the staff of Center for the Biology of Natural Systems, Washington University, St. Louis, MO 63130.

11. Jeavons, Research Report Summary.

12. The Wedell Group: Research Architects, *Intensive Small Farms and the Urban Fringe,* 2300 Bridgeway, Sausalito, CA 94965, 1976, 96 pp. "An agricultural land census by the Economic Research Service of USDA showed this year that while 400 million acres of cropland remained on the rolls as it has since figures were first developed in the early 1900's, over 110 million acres are in *declining production.* In addition production increases anticipated when agricultural restrictions were removed in 1974 and 1975 have not materialized as expected. . . . An extrapolation of ABAG [Association of Bay Area Governments] open space figures nationwide equals ap-

proximately 100 million acres of potential cropland available in the urban fringe or approximately equal to the croplands in current declining agricultural production. Stated another way, *urban fringe lands equal one fourth to one third of all cropland in the U.S., the world's richest agricultural country.*"

Pimentel, et al., "Land Degradation." "Each year . . . more than 2.5 million acres of arable cropland [are] lost to highways, urbanization, and other special uses [in the United States]. Approximately 40 million acres of land in the U.S. have been converted to urban uses to day; about half of this land formerly had been cropland. Other land uses have taken their toll as well. About 32 million acres have been covered by highways and roads so far. The passenger car system accounts for 50% of the total loss to highways, urbanization and other special uses" (p. 149). "Approximately 740 million acres are in pasture and rangeland" (p. 149). Of this "another 90 million acres of land . . . could be used for crops" (p. 153). (This last figure is undoubtedly higher if intensive techniques are being considered.)

13. Bolton Hall, *Three Acres and Liberty,* New York: Macmillan, 1918, 276 pp. The quotation from Lincoln appears on p. 231.

14. Perelman, "Efficiency and Agriculture," p. 1.

15. Sterling Wortman, "Food and Agriculture," *Scientific American,* September, 1976.

16. Perelman, "Efficiency and Agriculture," p. 5.

17. Ibid., pp. 40 ff.

18. John Jeavons, personal communication, February, 1977.

19. Carter and Dale, *Topsoil and Civilization,* p. 159: "Western Europe has a fairly sustainable agricultural system due primarily to the adaptation of its agriculture to the climate. Gentle misty rains instead of torrential downpours, and snow cover in the winter in combination with manuring, green manuring, crop rotation and other sound farming practices. . . . It should be emphasized that the present farming system of Western Europe was evolved primarily by farmers who owned the land they tilled. Little progress was made toward soil conservation during the Middle Ages; when feudalism and its communal system of agriculture offered little inducement for conservation and little was practiced. Crop rotation, manuring, liming, and other soil building methods became common only after feudalism and communal agriculture disappeared."

20. Perelman, "Efficiency and Agriculture," p. 9.

21. Revelle, "Resources Available," p. 170.

22. Perelman, "Efficiency and Agriculture," p. 12.

23. John Jeavons, personal communication, February, 1977. Evaluation based on a five-year Cornell University study completed in 1970 by Dr. Richard B. Root, an entomologist, and Dr. Jorma O. Tahvanainen, a Finnish scientist (see Jeff Cox, "The Technique That Halves Your Insect Popula-

tion," *Organic Gardening and Farming,* May, 1973, pp. 103–104); and on a U.S. Department of Agriculture estimate "that, even if all pesticides were eliminated, crop loss due to pests (insects, pathogens, weeds, mammals, and birds) would rise only about 7 percentage points, from 33.6% to 40.7%" (see Frances Moore Lappé and Joseph Collins, "Food First," Institute for Food Development Policy, P.O. Box 57, Hastings-on Hudson, New York, NY 10706, 13 pp., p. 7).

24. Howard, *Agricultural Testament,* pp. 160 ff.: "The crops grown by the cultivators in the neighborhood of Pusa were remarkably free from pests of all kinds; such things as insecticides and fungicides found no place in this ancient system of agriculture. . . . Insects and fungi are not the real cause of plant diseases but only attack unsuitable varieties of crops imperfectly grown. Their true role is that of censors for pointing out the crops that are improperly nourished and so keeping our agriculture up to the mark. . . . At Pusa, during the years 1910–1924, outbreaks of plant diseases were rare . . . poor soil aeration has always encouraged diseases at Pusa. . . . The unit species of *Lathyrus sativus* provided a good example of the relation between aeration and insect attack. . . . Surface root types were always immune to green fly while deep rooted types were always heavily infected. Types with intermediate root systems were moderately infected. . . . These sets of cultures were grown side by side year after year in small oblong plots about ten feet wide. The green fly infection repeated itself each year and was determined not by the presence of the parasite, but by the root development of the host. Obviously the host had to be in a certain condition before infection could take place. The insect, therefore, was not the cause but the consequence of something else." When Howard first came to Pusa much of the tobacco being grown was malformed and diseased, due to a virus. "When care was devoted to the details of growing tobacco seed, to the raising of seedlings in the nurseries, to transplanting and general soil management, the virus disease disappeared altogether. It was very common for the first three years, it then became infrequent; and between 1910 and 1924 I never saw a single case. Nothing was done in the way of prevention beyond good farming methods and the building up of a fertile soil."

25. Revelle, "Resources Available," p. 168: "In principle, however, most—perhaps all—of the energy needed in modern high yielding agriculture could be provided by the farmers themselves. For every ton of cereal grain there are one to two tons of humanly inedible crop residues with an energy content considerably greater than the food energy in the grain. If only 1/2 of this energy could be recovered by the fermentative production of methane or alcohol, the energy requirements of modern agriculture, including energy for production of chemical fertilizers, could be fully satisfied."

Gary Soucie, "How You Gonna Keep It Down on the Farm," *Audubon,* September, 1972, pp. 112–115. American agriculture generates each year ten times the solid wastes of all cities, towns, and suburbs combined—2.5

billion tons per year, of which animal wastes are 1.8 billion tons. "If this manure were divided up equally, every family would receive 25 cubic yards each year as their fair share. This would be enough to fill the average 9 X 15 foot living room to a depth of 5 feet or about chin level of most adults."

26. Jeavons, Research Report Summary.

Ralph Engelken has been growing corn organically for 19 years. His corn currently has 12% protein content with yields ranging from 100–150 bushels per acre average. (The 1971 U.S. average was 87 bushels per acre with an average protein content of about 9%. See Pimentel, et al., "Food Production.").

The Prospects for Small-Scale Farming in an Industrial Society: A Critical Appraisal of *Small Is Beautiful*

E. Phillip LeVeen

> We know too much about ecology today to have any excuse for the many abuses that are currently going on in the management of the land, in the management of animals, in food storage, food processing, and in heedless urbanization. If we permit them, this is not due to poverty; it is due to the fact that, as a society, we have no firm basis of belief in any metaeconomic values, and when there is no such belief the economic calculus takes over. This is quite inevitable. . . . Nature abhors a vacuum, and when the available, "spiritual space" is not filled by some higher motivation, then it will be filled by something lower—by the small, mean, calculating attitude of life which is rationalized in the economic calculus. . . .
>
> If we could return to a generous recognition of metaeconomic values, our landscapes would become healthy and beautiful again and our people would regain the dignity of man, who knows himself as higher than the animals but never forgets the noblesse oblige.
>
> E. F. Schumacher, *Small Is Beautiful*
> (Harper Torchbook, 1973, p. 109)

In light of the increasing scarcity of energy resources, it appears that American agricultural "efficiency" is built upon a fragile foundation. E. F. Schumacher and many others have called attention to the increasing dependence of our agriculture on scarce fossil fuel resources as a result of the rapid substitution of mechanical and chemical technologies for human and animal resources. This substitution is seen as damaging not only in its impact on scarce natural resources and environmental quality, but also in its effects on those individuals and their communities forced to undertake major transformations as a result of labor-displacing technologies. It thus seems reasonable to many that we should return to a small-scale agriculture which can serve to reduce our use of non-renewable natural resources, increase the quality of our environment, and provide more creative work for our citizens.

Dr. Schumacher goes further and believes that changing the scale of agriculture and using intermediate technologies will have important beneficial effects on the overall society. It will be infused with a new awareness of the limits nature imposes on human society and therefore be induced to accept the values of simplicity, nonviolence, and the dignity of the individual. Given such a conversion, the development of a low-consumption life style depends mainly on the creation of technologies more conducive to small-scale enterprise. It requires the implementation of more reasonable principles to humanize the operation of large-scale organizations, and changes in ownership patterns such as collectivized control over firms by employees or nationalization of very large firms. The transformation of agricultural production is a vital first step in the overall solution.

Dr. Schumacher continually argues throughout his book that the primary obstacle to the implementation of intermediate technologies and the development of human-sized institutions is the lack of belief in appropriate values. He lays much of the blame on the discipline of economics, which has promoted the idea of economic growth and the associated values of greed and avarice, all of which have created a materialistic and destructive life style. It will be argued in this paper that Mr. Schumacher gives too little attention to structural and institutional factors in his analysis of the reasons for our present predicament, and, because of this, he does not have a clear understanding of the forces which give rise to growth. Without an understanding of such forces, it is impossible to describe the types of social, political, and economic changes which must occur if we are to achieve the noble goals he sets forth.

To illustrate this point, we shall examine the major structural and institutional determinants of the adoption of labor-saving technology. This process is at the heart of agricultural development in the United States and in almost all other countries and is closely associated with most of the negative aspects of this development. The examples are drawn from the experience of capitalist American agriculture, but they are representative of more general processes at work in a variety of non-centrally-planned industrial systems ranging from capitalist to market-socialist economies.

There are three parts to the remainder of this paper. First, we shall examine the role of competition in the context of family farm agriculture in order to understand one set of reasons why farmers are forced to substitute capital for labor. Second, we shall examine the structural features of an advanced industrial agriculture, as found in California, in order to show how the dependence upon hired labor has created powerful incentives for farmers to employ labor-saving technology. Third, we shall show, in the context of California agriculture, why organized labor and farm owners both have a stake in the continued substitution of capital for labor. Here we shall suggest why, in a capitalist system, growth of this form is a necessary condition for social stability. From these examples will emerge a clearer notion as to why intermediate technology and work humanization are not finding ready acceptance, despite the present interest in developing these important ideas. This resistance arises not because of any lack of "meta-economic" values, but rather, from the inherent features of an industrial society.

THE FAMILY FARM AND THE TREADMILL

In many ways, the midwestern family farm resembles Dr. Schumacher's ideal production unit. Fifty years ago, the typical farm was small, owned and operated by a family, and not particularly dependent upon mechanical technology. Farmers then used neither inorganic fertilizers nor pesticides. Today, the family farm has become heavily dependent on both machines and fertilizers; it has grown so large that soon most families will not be able to operate it with their own labor. And, even fewer will be able to own such a farm. Has this development occurred because of the wrong values, or is there something else going on that may be less apparent, even to the participants in this process? Let us look at the process economists call the "treadmill."

The family farmer is part of a process over which no individual or institution has control: it is the process of economic competition. When the farmer stopped producing primarily for his family and began to produce for the market, he became involved in an inescapable treadmill. Farmers were not in the habit of reading Keynes (who was not even around during the early years of this process), but they did learn how to stay alive. To do this, they found it necessary to use bigger and bigger machines. To understand their situation, the following hypothetical and simplified example should help.

Suppose a new harvester is developed which permits the farmer to extend his family's labor over more acres of cropland. Suppose only one farmer finds out about this device and is able to purchase it before all his competitors do, and that the farmer also has enough unused land to place under cultivation so that the harvester can be used to full capacity. This farmer will be able to increase the volume of crops produced by his family's labor and, as long as the incremental value of this increased production is more than the annual cost of the harvester, the farmer's income increases. This type of innovation is called "labor saving" because it reduces the amount of labor necessary to produce a given amount of the crop.

Now, suppose that the farmer's competitors, seeing the advantage of the harvester, attempt to purchase similar machines. Two things happen. First, in order to derive the harvester's cost-reducing advantages, any farmer who buys the machine must be able to increase the size of his acreage. If there is excess cropland, then all farmers can expand. But, as is more likely the case, if land is already farmed to capacity, expansion of some farms can occur only if other farmers sell off some of their land. Thus, only those farmers who can buy the additional land can profitably employ the machine. As a result, the machine eventually displaces some farmers who sell their land to the expanding farmers. Other farmers who cannot expand continue using the previous technologies at a substantial cost.

The second effect of the spread of the harvester's adoption is the reduction of farm production costs, which fall by the amount of the cost savings permitted by the use of the new harvester. Since farming is extremely competitive, lower production costs mean lower prices in the market; and, because of an inelastic demand for food, lower market prices mean a reduction in total gross income. As prices fall, the benefits derived by the first farmer who adopted the harvester begin to disappear. For the farmers who adopted the harvester later, the lower prices wipe out the expected profits. They are no better off after they increase their farm size than they were before the process started. For the farmers who were unable to grow, the falling prices mean lower incomes because they do not have offsetting reductions in their production costs. That is, non-adopters experience a substantial economic penalty for their failure to keep up with the latest technology, while the rest of the farmers are, in the long run, no better off for their expansion.*

In the face of an ever-increasing number of such labor-saving developments, the farmer soon learns that he is in a race for his survival, especially once he has a debt which he must service. If he is not among the first to adopt some innovation, he will be penalized. However, this race can have only a limited number of winners. Given the demand for food and the limited availability of land, not all farmers can expand. Once a farmer begins to fall behind, he will have ever-increasing difficulty in catching up. The successful farmer has the profits and access to the financial capital necessary to underwrite the next round of expansion and investment, but the majority of farmers do not. Even though only a small fraction of farmers can keep up with the latest developments, their production is a large portion of the total crop; thus, their impact on industrywide costs is profound.

Under this competitive process, the initial egalitarian patterns of land distribution and farm income have been transformed into a highly stratified system in which a few large farmers control most of the farm income and land, while a large number of farmers fight for the remains. But even the most successful farms are not assured of continuing success or even of receiving a return comparable to what an industrialist might earn on his investment. The old clichés such as "You have to move forward—you cannot stand still and survive," have a very real meaning to all farmers. They are not, as Dr. Schumacher suggests, outmoded ideas left over from the nineteenth century.

To illustrate this point in another way, suppose that a farmers' organization is formed and all farmers are forced to join. Assume that this organization is given governmental sanction to enforce the following rule: No farmer can adopt a new technology without the consent of all other farmers; and all requests to adopt a new tech-

*This analysis is, of course, overly simplified. During the years when government commodity programs effectively stabilized prices on the basis of "parity," the treadmill worked through the land market instead of directly through market prices. The impact of competition, however, was essentially the same, as land prices rose to drive out farmers.

nology must be accompanied by an analysis of the overall impact of the technology on the entire farming community, given its widespread adoption. Since most labor-saving technologies do not in the long run benefit farmers, it is reasonable to expect that the rate of substitution of capital for labor will be drastically reduced by such a decision rule. Farmers would be forced to take a totally different perspective on the processes which now stimulate each individual farmer to disregard the impact of his behavior on the welfare of others. This is not to say that some substitution of capital for labor will never occur; it does mean that new machines will not be introduced at a rate faster than farmers are willing to leave the land voluntarily. It also means that farmers will be more willing to purchase non-labor-displacing equipment which simply makes farm work easier or less costly. It also means that farmers will be willing to consider technologies which increase the yields of their lands through techniques which intensify farming subject to the demand for any increased output.*

The implication of this example for the development and utilization of intermediate technologies should be self-evident. The treadmill process sensitizes each farmer to be receptive to technologies which reduce his costs below those of his competitors. If a firm supplying farm equipment wants to ensure its own success, it too

*The decision rule we have proposed above is equivalent to granting farmers monopoly power vis-à-vis other sectors of the economy. This power is used to benefit farmers, but it has the following negative impacts on the rest of the economy insofar as it reduces the rate of substitution of capital for labor.

Farm equipment producers would produce fewer machines and hence earn lower incomes. Employment in this sector would be reduced. Landowners, such as railroads, would earn lower capital gains since labor-saving technology increases the demand for land in comparison with yield-increasing technologies. Urban industrialists would have to pay higher wages because more farmers would remain on the land instead of coming to the cities to compete in urban labor markets.

There is also a question of whether food prices would be higher without labor-saving technology. Since such technologies place farmers in a cost-price squeeze, those farmers who find their incomes declining may attempt to compensate for the lower prices by producing more food which helps to increase food surpluses and lower prices. Yield-increasing technologies have the propensities to create surpluses as well; so unless farmers can jointly control production, the decision rule does not necessarily mean an end to surpluses. If farmers use their power to control output, they will keep prices higher than they have been under the present system.

must compete with other firms and must compete for capital in order to succeed; it must exploit the farmer's vulnerability to the treadmill. Should a firm decide to develop small-scaled, non-labor-displacing technologies, few farmers—including those who might otherwise want such machines—can afford to be interested because of the implicit penalty involved.

Some proponents of small-scale farming have noted the high degree of economic concentration in the farm machinery sector and have suggested a connection between this monopoly power and the type of technology designed and produced. Such an argument assumes that smaller and more numerous farm machinery suppliers would have to be more responsive to the "needs" of farmers and thus would produce intermediate technologies. While there is no denying that monopoly power in this industry has had some impact on the types of equipment offered farmers and, more important, on the profits of machinery producers, it is not reasonable to think that more competitive firms would operate differently. The logic of competition would force the farmer to demand, and the machinery producer to want to supply, labor-displacing equipment. Only if the farmer cannot improve his income—or thinks he cannot improve his income—will he stop looking for bigger machines.

Here, then, we have an explicit example of a structural feature inherent in the competitive market system. It explains the irrational—from the overall social viewpoint and from the perspective of farmers taken as a group—patterns of human and material resource development. This pattern derives not from the particular attitudes or values of any individual, but from the fact that each person is placed in a competitive relationship with all others which forces the particular type of behavior. This behavior is entirely rational and necessary—indeed, the farmer's survival is at stake. The design of other technologies, even if they can be shown to be technically feasible, will have little impact unless they fundamentally alter the farmer's competitive position or unless they are accompanied by changes which stop the treadmill. The treadmill could be interrupted in the future by changes in the relative prices of energy and land in relationship to labor, which will make further mechanization less profitable. However, even if this eventuality should come soon, our system will still be heavily dependent upon energy resources until technology becomes available which allows farmers to move in a new direction. Thus, in the future it is possible that the intermediate technologies, which now are given little attention, will find a home in our family farm agriculture.

But the relative prices of energy and land may have to change by a substantial amount before they influence California industrial farmers to move away from labor-displacing mechanization. To understand why, we now examine the structural features of this system of production.

INDUSTRIAL FARMING: THE IMPORTANCE OF HIERARCHICAL CONTROL

While the large scale of California agriculture is the most obvious difference from midwestern family farm agriculture, the difference in the internal organization of the two systems is of greater interest to this analysis. The typical farm in California is difficult to describe because of the great diversity of crops grown, but one structural feature of most California farms which is substantially different from the midwestern family farm is the importance of hired labor. Most family farms depend mainly upon the family's labor; in California, most farms depend upon the owner to provide management decisions and leave the manual field work to hired workers. Of course, there are exceptions to this, and some farmers do provide substantial amounts of labor as well as management. But the majority of the production in California occurs on farms in which ownership and management are separated from labor. On many of the largest farms, ownership and management functions are separated as well, creating a production system which is structurally equivalent to that which prevails throughout the industrial economy.

Of concern to this analysis is the develoment of a dependence upon hired labor in California and the associated development of a hierarchical pattern of authority which gives to the owner and manager both the power and the incentives to make certain kinds of decisions in the choice of new technologies. This structural differentiation in the roles of labor and management, which is not present in the family farm system, introduces a new reason for the adoption of labor-saving technology and the growth of farm size which is not found in the treadmill analysis. To understand this, we must quickly review the reasons why family farmers were not able to establish a foothold in California.

Large-sized farms are not a recent development in California. In 1920, fully 30 percent of all harvested cropland was located on farms of more than 1,000 acres (for comparison, less than 0.5 percent of Iowa cropland is so farmed at the present time). Thus, the large California farm predates the development of sophisticated mechanical technologies, which means that in contrast to the Midwest, farms did not grow larger in order to accommodate bigger machines. In fact, the reverse may be true; machinery became larger to support the large farm. This raises the question as to why California farms became large in the first place. The answer to this is to be found in the initial land distribution patterns which allowed large holdings to be amassed by a few individuals (the Midwest was settled in 40-, 80-, and 160-acre homesteads) and, most important, in the development of a captive and, therefore, cheap supply of agricultural workers to provide the labor necessary to farm these large holdings. In the Midwest, labor was expensive and scarce; farmers were forced to rely upon their own labor.

When the family farmer tried to settle in California and establish a midwestern-style farming system, he found himself competing with farmers who employed large crews of low-paid workers. The first farm workers were Chinese who had finished working the transcontinental railroads, but later other minorities took their place. To the farmer who used these workers, labor became a "variable" cost of production. When the worker was not needed, he was "let go" and ceased to be an economic cost to the farmer. Migrating workers could find work on other farms producing different crops, but as a group, farm workers historically have been both low paid—relative to urban workers—and underemployed. The family farmer, who planned on using his own labor, was thus forced to compete with a labor system which kept farm commodity prices low. As a result, such a farmer, with his fixed labor supply, could not survive. Family farmers who were able to obtain land were therefore forced to depend on hired, temporary workers. Thus, the scale of farming was no longer confined by the availability of family labor; and it was possible, even without mechanization, to maintain very large-sized farms. Once these patterns of farming were established, an economic interest in their perpetuation developed; and a variety of supportive institutions, both public and private, were created to ensure this goal.

The maintenance of an adequate supply of low-wage labor has

been an ever-present concern of California farmers. Should this hired labor force be lost, farmers would be forced to compete in urban labor markets to find the necessary labor to maintain their operations, which would quickly eliminate the differential in wages and working conditions between the two sectors. Higher wages would mean higher production costs, and therefore, increased prices and lower demand, especially for fruits and vegetables. But of greater interest to this analysis is that higher wages would weaken the basis of large-scale, industrial agriculture relative to the smaller family farm, since the family farmer's labor would become worth more as a result of higher wages paid hired labor. Thus, low wages in agriculture have been important to the maintenance of a large-scale agricultural system, and any major interference with this labor system has been regarded with particular hostility by those who have benefited greatly from it.

It comes as no surprise that California farmers have led the nation in the adoption of labor-saving technologies. Not only does mechanization help to reduce the critical dependence upon an uncertain hired labor system but also, through its displacement of workers, it helps to maintain an abundant supply of labor for the remaining agricultural employment and thereby keeps wages low. Displaced workers have nowhere else to go.* Mechanization also changes the nature of the labor force which remains in the mechanized sector. This labor force is used to running and maintaining complex equipment and is more like the industrial work force in terms of its skills. Should farmers be forced to pay higher wages in the future, it will be easier to do so with this type of productive labor force. We shall develop this theme in the next section.

Thus, the California farmer is committed to labor-saving technologies for two reasons. The first is his competition with other farms and his desire to succeed in the treadmill. Secondly, the farmer is aware of his vulnerability to an uncertain labor situation, and, if given a choice, knows that the machine will not strike, ask

*Obviously, an important aspect of this analysis concerns how farm workers are segmented from the rest of the labor force to prevent an equalization of the urban and rural wages. Without going into detail, it should be pointed out that, except for a short period in the 1930's, farm workers have all been largely first-generation immigrants or visitors who lack language and other skills necessary in the urban market. Racial discrimination has also been an extremely powerful segmenting device.

for higher wages, or fail to appear at some time in the future. Thus, even when a machine does not lower production costs, the farmer may still have considerable incentive to use it.*

This comparison between the California and the midwestern examples reveals an important development resulting from the increasing specialization which is part of the maturation of an industrial economy. This development is the separation of the functions of enterprise into three different sets of activities—ownership, management, and labor—carried out by at least two different groups of individuals. In the family farm, all three functions are united in one family; and although the family farmer increasingly must share decision-making functions with the banker or other external groups, the basic conflict between ownership of the farm's assets and the development of its labor resources is resolved by the individual farmer. He represents both sides of the conflict and can best balance the conflicting needs of higher returns on investment and higher returns on labor. In this case, labor has an equal, or at least a more equal claim in the development of the farm enterprise than it does in California, in which the different functions of labor and management are separated, and conflicts are resolved through the hierarchical structure of authority which gives the farm owner complete freedom to make production decisions. Even though such decisions directly affect the farm workers, farm labor has been unable to exercise much influence in the decision-making process. Thus, in California, labor has a more unequal claim on the development of the farm enterprise than in the family farm system.

It is instructive to notice that the California farmer, even if forced into a collective organization and required to have the consent of other farmers before adopting a labor-saving device, would—in

---

*Interestingly, we might foresee the receptivity of industrial agriculture to non-chemical production technologies, especially if these alternatives do not increase dependence on the traditional labor markets. For example, integrated pest management systems are based on the use of skilled consultants. These services reduce the use of pesticides in those crops for which systems have been developed; if anything, these systems are supportive of industrial agriculture because the cost of consultants decreases with the size of farm. Thus, as pesticide costs rise, it is likely that the larger farms will have some advantage over the smaller ones unless a new type of delivery system is evolved. In short, not all energy- and fossil fuel–saving technologies will be resisted—only those which challenge the basic relationship of capital and labor.

contrast to the family farmer—continue to want such technology. In the family farm case, the collective organization permitted farmers to stop the introduction of labor-saving technology because they could see that their actions led eventually to the displacement of their own labor. In California, such technology does not threaten the farm owner's labor, so he is free to follow a course which is detrimental to labor's welfare. If, on the other hand, the decision-making power in California were given to labor, a very different reaction toward technology could be anticipated. Machines which displaced workers would be acceptable only insofar as additional employment could be created for the displaced worker.* This example thus illustrates how the separation of farm ownership from farm labor functions, together with the existing structure of hierarchical control which reduces labor's power to influence key decisions, creates an incentive to choose labor-saving technology. Because this system confers considerable benefits on owners at the expense of farm workers and because farm owners owe their very existence to their ability to control the labor market, it is highly unlikely that reforms of the type Dr. Schumacher suggests (for example, collective ownership) will make many inroads in California agriculture. Such experiments threaten the very core of the present system's existence.

Before moving on to the next section, there is one other important development in California agriculture which has bearing on the prospects for small-scale production units. This concerns the increasing control of agricultural production by agribusiness. There have been many studies of the economies of size in California farming. Those relating to the production of fruits and vegetable crops indicate that the small farm has very little cost disadvantage in comparison with the larger farms. Many of our fruit farms are, in fact, relatively small in terms of acreage. Most of our vegetable

---

*This example abstracts from another interesting aspect of agricultural labor in California, namely, the connection of many workers with Mexico and the increasing use of illegal aliens in California. Workers thus may have conflicting interests. Those who reside in the United States and who want to keep others out of their labor market may have different concerns from those who work for part of the year in California and spend their earnings in Mexico. Mechanization may be more threatening to the illegal alien and to the casual worker or the migrant farm worker than it is to the stable, United States–based worker.

are very large in terms of their acreage. Both types of farming are labor-intensive, at least as of this moment. There is much interest in mechanization in these types of farming as well, and new devices are being developed rapidly.

It is not feasible for an individual family to compete with a large farm, simply because of the very limited amount of land which a family using only its own labor can hope to harvest in fruits and vegetable crops. The family cannot afford the equipment for this small amount of land; and, increasingly, the volume of output a single small farm can produce is not sufficient to allow it access to the major commercial markets. Perhaps the major reason for the success of large vegetable farms is their ability to deal with financial institutions, processors, and food distributors. The increasing scale of vegetable farms derives not from their more efficient use of resources over the smaller farm, but because of their ability to relate to the needs of the increasingly concentrated set of institutions called agribusiness.

The lack of major economies of size in fruit and vegetable production and the existence of advantages associated with access to credit, supplies, and markets have led many smaller farmers to form cooperative associations through which some of these advantages may be made available to the smaller unit. Those interested in agrarian reform in California have chosen to develop labor-intensive farming projects as a means of incorporating the individual family into the agricultural system in a new way. For example, there have been several recent innovative efforts to extend the concept of cooperation to include—in addition to the traditional supply and marketing activities—new forms of collective production. A few notable successes in labor-intensive cooperative farming suggest the possibility that non-traditional farming organizations, built on far greater integration and the sharing of all aspects of farming, will find a place in California and will offer an alternative to industrial farming organization. This type of farming is closer to the family farm ideal because it is based on a sharing of all economic functions—ownership, management, and labor—by all members in a democratic, non-hierarchical, decision-making context. However, these successes should not be taken as a sign that small-scale farming will be able to effectively compete in a wide variety of crops, for such cooperatives have a very tenuous existence. Cooperation is time-consuming and costly; therefore, the production cooperative has a difficult time competing with a well organized industrial farm and maintaining its basic cooperative structure.* But cooperatives have more important problems they must also confront.

Even if small farmers are successful in mastering the problems of sharing equipment and marketing—as long as their activities are confined to the labor-intensive vegetable and fruit crops—they must contend with the same competitive forces which we analyzed in the family farm case; and this competition leads to dramatic swings in the prices of vegetable crops. Since the markets are limited, small changes in production can make the difference between boom and bust. Small farmers cannot tolerate bust—they do not have the financial resources to do so. For example, near Fresno, four families leased four acres and grew cherry tomatoes from which they jointly grossed $85,000 in the first year. The next year several other farmers, hearing of this success, tried—with the help of community development funds—to duplicate the results. However, what made sense for a few families did not make sense for several families. The market was flooded, to everyone's detriment. Clearly, an institution is required to prevent such lack of coordination; and, in order for this to work, all farmers would have to agree to make collective production decisions. Few established farmers would be willing to help out new farmers, especially if it means reducing their own production; thus, the necessary cooperation is unlikely, given the present realities of a highly stratified agricultural community. Small farmers, moreover, aren't particularly welcomed by the existing marketing orders. Grower marketing cooperatives which have helped established farmers to gain a measure of control over their markets, have increased the price stability in many of the fruit and vegetable crops because these growers feel threatened by such experiments.

The small-scale, labor-intensive farm poses two threats to the

---

*The biases against such non-hierarchical decision-making processes inherent in a competitive economy are important constraints on the power of any enterprise, no matter what the motivation of its owners, to evolve less alienating forms of worker control. Cooperation and democratic processes are not "efficient" in the context of such a system even though they may be essential to the creation of a truly non-alienating workplace. Thus, the particular problems of the cooperative farm are illustrative of the more general limits to the effectiveness of work humanization schemes throughout the economy.

existing system of production. If this type of farm grows in number and importance, it will increasingly threaten the incomes of larger farmers and the stability of the marketplace with the additional output. Perhaps equally important, small farms threaten the labor supplies of the existing farms because they provide an alternative method of organization outside the hired labor market. Many of the experiments involve the participation of farm workers who otherwise would have remained in wage labor. Therefore, it comes as no surprise that California agriculture has not welcomed these experiments and that the public and private support institutions, whose goals have been defined by the existing system, are not anxious to help these groups get started. Finally, because these small farms must continue to compete with the existing structure of agriculture, the return to labor is still determined by the hired labor wage rates which continue to be well below those of urban workers. In other words, the benefits of such farming are not overwhelming to those who must do the labor. Unless the rest of the labor force in agriculture increases its wages, the small farm experiments will continue to face uncertainties in their ability to survive in Caifornia commercial agriculture.

One final point must be made concerning small farms in California. Because of changes in the requirements of the processing and food distribution system, farmers have increasingly sacrificed quality and flavor in their effort to produce with labor-saving technologies for a national and international market. This leaves a significant unfilled demand for high-quality fresh fruits and vegetables. For the energetic farmer who is willing to develop an alternative market network with urban markets, the possibility of truck-farming vegetables and fresh fruits using organic and labor-intensive methods may be increasingly bright, especially in view of the importance of energy in the transportation of food to the market. Here is a niche in the present system which could well be filled with unorthodox farm organization and operation, and here is one possible source of immediate interest in intermediate technologies.

In summary, we have identified in California's industrial agriculture a variety of economic and political processes which have combined to force the individual farmer to grow ever larger and to substitute capital for labor. These processes include—in addition to the competitive treadmill outlined in the first section—the development of a deep and antagonistic relationship between farm owner and farm worker, the lack of farm-worker power to influence basic production decisions, and the increasing concentration in the private corporations which provide the farmer his credit and his access to profitable markets. Added to these must be the structuring of public-support institutions and government regulations to serve the largest farmers, and the emergence of powerful political interests which protect the large farmer from all efforts to change existing political-economic relationships. Thus, if small farms are to make any major impact on California agriculture, far more than the development of intermediate technologies or the espousal of the ideals of smallness and non-violence will be required. Indeed, unless a major shift in the relationships between the farm worker and the farmer can be achieved, small-scale agricultural development—if it succeeds at all—will be confined to the marginal role of exploiting the markets in which agribusiness is no longer interested.

THE FUNCTIONAL IMPORTANCE OF CAPITAL-INTENSIVE GROWTH IN AN INDUSTRIAL ECONOMY

In recent years, mechanization has virtually eliminated much of the unskilled labor which used to be required for the maintenance of many industrial farms. While this has been particularly true in the production of field crops, increasingly, the processed vegetable crops are also being mechanized. While there remains a large need for manual labor in the fruit and fresh vegetable crops, the tendency for even these crops to be mechanized suggests that California growers may soon eliminate their historic dependence on cheap, unskilled wage labor, and instead will require a more skilled and more permanent labor force to run the industrial farms. If this occurs, all the conditions of urban industry will be duplicated in agriculture, including the characteristics of the labor force. This development will mean that higher wages will no longer threaten the basis of the land-tenure system.* Mechanized agriculture reduces the labor share of total costs and makes the individual worker

*It might be argued that, in view of economies-of-size studies which reveal only slight advantages of the highly mechanized industrial farm in relationship to the equally mechanized family-run farm, large farms will still be threatened by smaller units. It is quite true that the industrial farm does not enjoy the same relative advantage vis-à-vis the smaller farm that General Motors enjoys over the family automobile business, but it must

so much more productive that he can be paid the higher wages without penalizing the employer. In other words, the traditional resistance to higher wages for farm workers may be broken down by the continuing drive toward mechanization. However, should mechanization be stopped, the farmer will not be willing to increase wages, for to do so would be to reduce his own welfare. Only when the farmer can "share" the benefits of his workers' increased productivity, or when he can force the consumer to pay the additional costs, will it be possible for agricultural wages to continue rising. The farmer faces a highly competitive market. This competition is not only within California, but between California and other regions of the world. Thus, the farmer cannot pass on higher production costs without serious impact on his relative market position. Should production costs grow too high, it is possible that our supplies of specialty crops will not be grown in California and will, instead, come from cheap labor areas in Latin America or other regions of the world. Thus, labor-saving technology, which keeps farm production costs low, becomes a key ingredient to any development of a new relationship between farmer and farm worker.

The increasing power of labor unions in California suggests that the above scenario is increasingly likely. If the farm-worker unions develop as urban labor unions have done, they too will agree to the strategy of labor-saving technology as long as they share in its benefits. The long-run effect of this continuing substitution of capital for labor, especially in agriculture with its limited potential for market expansion, will be a dramatic reduction in the total labor force in agriculture. Elsewhere, unions have accepted this long-run condition, as long as there are safeguards for existing employees,

---

be remembered that the family farm capable of competing with the large industrial farm is also a large investment. A farm of this type could easily cost $500,000 which means that, unless the family already owns the land, a substantial barrier exists for any prospective farmer who might want to take advantage of the larger farm's vulnerability. This barrier will therefore provide an ample protection to the larger farm, even if its wages were to climb to parity with industrial workers. The only possible exception of this argument is in the Westlands and other federal water districts in which large farmers may be forced to sell their lands at less than half market price. Here, smaller family famers might be able to secure the necessary capital to overcome the barrier and establish themselves in competition with larger farms.

because higher wages in the short run are necessary to the internal stability of the union organization and because these are the only conditions under which employers will agree to higher wages. Employers are less likely to starve than are the strikers if there is a prolonged strike.

What this argument suggests is a general proposition that labor-saving technology is the one common point at which labor and capital can meet, given the hierarchical nature of control and the lack of employers' willingness to reduce their present level of welfare. Labor-saving technology moderates the potential conflict between the two groups by providing enough benefits to satisfy the short-term interests of both. In this way, the potential conflict which emerges as a result of the stratification of economic functions in a mature industrial society can be deferred through the process of growth. This is why such technology can be said to provide a vital stabilizing influence in industrial society.

The question arises, however, as to whether this process of growth can continue indefinitely. For reasons outlined above, if growth requires the continued exploitation of scarce natural resources, it must be eventually limited. The process has another contradiction built into it. With the continued substitution of capital for labor throughout the economy, more and more individuals are being forced to look for work in the marginal, low-wage sectors of the economy or simply to move onto the welfare and unemployment rolls. Thus, the important stabilizing impact of the labor-saving growth on the relationships between labor and capital may be forced to come to an end when labor unions find themselves with very small memberships and when the cost of maintaining the unemployed workers place severe burdens on the entire economy. Labor cannot indefinitely sacrifice its long-term position so that its present members can have high-wage work. The question is—in view of the basic conflict between labor and capital—how can these contradictions be resolved in a mutually agreeable fashion?

For there to be an effective resolution to this long-term problem of maintaining social stability in the face of resource scarcity, environmental degradation, and worker alienation, the conflict between capital and labor must be directly confronted. As of now, when the possibility of growth is denied, the only way one group can improve its welfare is to reduce that of the other group. This is why one finds management and labor allied against environmentalists who increasingly threaten the basis of their past harmony. In order to move away from this situation, the worker must be convinced that adjustments to scarcity will not be asymmetrical, with labor bearing most of the costs. As long as decision-making in the firm is hierarchical, this belief will persist.

The family farm, removed from the treadmill, provides an ideal example of an institution which could accommodate an adjustment to a world of scarcity without major upheaval. The challenge is to find some mechanism which has the same effect but which works in the setting of a complex industrial organization. The Scott-Bader Corporation, described in *Small Is Beautiful,* is one possible example of a firm in which employees are integrated into the decision-making process. The cooperative farm is another example of complex production organization which can deal with the central problem of divergent interests between owners and workers. These types of development could make the transition to a lower-consumption life style more possible, especially if connected with institutional developments which remove the competitive drive for individual firm survival from the interaction between firms and industries. Unfortunately, we are dealing with exceedingly difficult reforms in describing such general changes; and there is little reason to expect that industrialists will voluntarily give up their present power without being forced by strong governmental pressure. The few efforts to "humanize" work reveal a general tendency for management to terminate any experiment in which workers gained sufficient control to threaten the prerogatives of management, and this includes experiments which had very positive impacts on worker productivity and on the profits of the business.* In other words, profits are less important than management control; and this suggests that Scott-Bader was indeed an unusual case.

To conclude, the processes of technological change and economic growth have played crucial roles in the determination of our present high-consumption life styles. In the preceding analysis we

---

*For a general review of the failure of work-humanization programs in several western economies, see: D. Jenkins, *Job Power, Blue and White Collar Democracy* (Garden City, N.Y.: Doubleday & Co., 1973); Lars Karlsson, "Industrial Democracy in Sweden," in Hunnius et al., *Workers' Control* (New York: Vintage Books, 1973); and Andrew Zimbalist, "The Limits of Work Humanization," *Review of Radical Political Economics,* vol. 7, no. 2 (Summer 1975), pp. 50–60.

have seen that the importance of these processes is more than the failure of Keynesian economics to deal with the "right" values and more than a problem of greed and avarice. The resolution of the crises, so well described in *Small Is Beautiful,* will depend on important structural reforms as well as on the creative development of new ways of living within our means. The exact nature of these reforms is impossible to specify, but a fundamental ingredient must be the development of new decision-making processes which give more equal weight to the welfare of all groups within society, and which allow each individual to understand how his or her welfare is related to that of the other groups. A just political process can make many of the other difficult decisions less threatening. Given such a setting, the "meta-economics" of Dr. Schumacher may become a possibility. The industrial world is the reality with which we must deal; we cannot recover the simplicity of the family business, nor can we use the ideal of the nineteenth century to build a new and better society. Understanding the structural and institutional origins of our present crises is a necessary first step if we are indeed going to make the industrial world into a place which is organized "as if people mattered."

Rationalization in Agricultural Development

Vashek Cervinka

The books about farming methods written in Central Europe in the first half of this century did not speak about the mechanization of agriculture, but rather about the rationalization of farm production. A dictionary may offer the following equivalents to the word rational: "pertaining to the reason, sane, logical, not absurd." Consequently, the rationalization of agriculture is reasonable and sane; however, mechanization for the sake of mechanization or modernization for the sake of modernization as we have witnessed in so many countries, may be considered illogical and absurd. I believe that we should be thinking again about the rationalization of agriculture and food production, and apply it both to the developing and developed countries. This paper is mainly related to the problems of production, productivity, labor, energy, and farm technology in developing countries.

A very costly but valuable experience has been obtained from both the successes and failures of agricultural projects in various developing countries. As a result, the optimism of the 1950's and 1960's has been replaced by a special hybrid of maturity, the skepticism and realism of the 1970's. There is a general consensus that no fast and simple solutions are available for agricultural progress in developing countries. No revolutionary changes are applicable and feasible to agricultural development; even the Green Revolution has not fulfilled all its high expectations.

Nearly everyone who worked on an agricultural development project has an identical experience of a conversion from an extreme idealism to a type of realism. I had an opportunity to work as an agricultural engineer in a tropical country during the 1960's. My training was in practical farming, under the European conditions of mechanized agriculture. I had knowledge of designing, testing, operating, and servicing tractors and farm equipment, as well as experience in the management of operations. However, I had absolutely no knowledge of the technological, social, and economic problems of tropical agriculture. Being basically optimistic, I expected that the technical team, of which I was a member, would easily train farmers to operate and service tractors and equipment, and assist in developing new land, and that the food and agricultural problems of a developing country would be substantially solved by way of farm mechanization.

After six years of my work, knowing that many farmers were trained in mechanized agriculture, and that thousands of acres of new land were cultivated, I was leaving the country, still optimistic but much less confident about the solutions to agricultural progress in a developing country.

One of the decisive moments affecting my experience was a demonstration of mechanized rice harvesting. Two combines were delivered to a village. These machines easily harvested about twenty hectares of rice. Nearly the whole population from the large village was standing idle at the edge of the field, curiously watching, and probably admiring, the working capacity of both combine harvesters. However, this type of experience helps one to realize that it may be completely absurd to import very expensive machines which take away work from farmers who have no industry, no services, absolutely no other place of going to find a new job. Basically, it was a classical situation when a capital-intensive technology is introduced in a labor-surplus society.*

Frequently a question has been asked about what type of model tropical agriculture should follow. Is there any model worth following? Can an experience of one country or region be applied to a completely different socio-economic environment? Can the type of low-labor and high-energy intensive farming be transplanted from California or Iowa to a developing country? Probably not. Should it be the Russian model? Rather not. What about a labor-intensive Chinese model? There are still so many unknowns about China, and if everything in China is as positive as described in numerous

*John Cole, *The Poor of the Earth,* Boulder, Colo.: Westview Press, 1976.

books, will the farmers of tropical regions be ready to work so hard and sacrifice as much as Chinese farmers have done? Neither Chinese farmers nor tropical farmers have ever achieved a higher material standard of living, but while the farmers in China have been struggling with nature for centuries, the rural population in the tropics has historically been living in relative harmony with nature. These factors have definitely affected the socio-economic and psychological environment in rural areas of both regions. Further, neither Russian nor Chinese models have proved that they would work in free societies, where advancement is rather desirable to encourage. Consequently, what model would be the closest one to the natural conditions of farming in the tropics, and easy to apply? Is there any farming model which is (or was at a given time) labor intensive, operating at relatively low inputs of fertilizers, chemicals, and machinery and achieving both a high energy efficiency and crop productivity per hectare of cultivated land?

Heichel indicated in his study that energy efficiency in corn production expressed by the ratios of food energy yield to cultural energy inputs, was higher in the Midwest in 1915 (4.8) and 1938 (4.9) than in 1969 (4.4) or in California during 1972 (2.2).* A similar relationship can also be found in various other studies. This trend of energy efficiency can be described by the mathematical models of inhibited growth. The problem is very well analyzed in the book by Myrdal, *The Challenge of World Poverty.*†

The farming in the Midwest from 1910 to 1940 had characteristics similar to the European farming technology which was, however, achieving higher yields per hectare of the cultivated land during the same period. As mentioned above, the agricultural literature was speaking about the rationalization of agriculture at that time. This type of farming was characterized by the utilization of both mechanical and animal power, simple equipment, labor-intensive operations, utilization of manure, compost, green manure, and crop residues. This maintained the soil fertility, relatively low application rates of fertilizers and chemicals, and relatively high productivity per unit of farm land. The mixed farming system was

---

*G. H. Heichel and C. R. Frink, "Anticipating the Energy Needs of American Agriculture," *Journal of Soil and Water Conservation,* vol. 30, no. 1, 1975.

†Gunnar Myrdal, *The Challenge of World Poverty,* New York: Pantheon Books, 1970.

prevailing at that time, the farmers working intensively both in the crop and animal production for the whole year. A higher proportion of the population was living in rural areas, since the manufacturing industry was in the initial stages of development. That was a time when land reforms were practically completed in Europe, the land being distributed from the large estates, owned by the aristocracy, to the individual farmers who cultivated the land.

A model of European farming from the period of 1910 to 1940, adjusted to the situation of the last quarter of this century, seems to be practical and rational for agriculture in the developing countries. This model may be in harmony with the present technological, social, and economic factors in the developing regions.

The application of the described model for agricultural development would be possible if the following conditions are fulfilled:

1. The land reform is completed and ownership given to the farmers who cultivate the land, and, thus, the equality of citizens in developing countries will be increased.
2. The agricultural education is oriented to practical farming needs (for example, the present price of rice combine harvesters may be equivalent to four hundred years of a farmer's income in some developing countries; consequently, it seems absurd to train farmers in the "modern" agricultural schools to operate and service combine harvesters).
3. The growth of grass-roots farming cooperatives is encouraged.
4. The governments in developed countries will realize that a long-range technologically, economically, and socially sound progress in developing countries is more relevant for their own economies than instant profits today.
5. Unrealistic dreams and goals will be replaced by the rational planning and management of agricultural development.

The scientific and technological knowledge is available for the process of agriculture in developing countries; it is a great challenge for man to apply it in a rational way.

Agriculture and Intermediate Technology

Michael Perelman

In spite of the natural advantages of soil and climate in the United States, this society has never been successful in using its bountiful resources. Figure 1 shows comparative food production data in various parts of the world. Nations such as Switzerland and Austria grow twice as much wheat per acre as the United States. Greece produces twice as much rice per acre, and Austria manages to produce twenty percent more corn per acre than the United States.

Large-scale energy production should also be faulted for excessive cost in terms of energy resources. For each pound of food to be delivered to United States tables, about thirty pounds of soil are eroded in the United States. The United States burns about ten calories of fossil fuel for each calorie of food produced. Now this figure involves a serious underestimation of the actual energy costs of United States farming practices. For example, modern energy-intensive farming is based on the use of manufactured inputs instead of products and processes produced on the farm. Manure gave way to manufactured fertilizer and tractors replaced horses. The work done by people who took care of the horses or applied the manure would generally be done a few yards away from home. A full accounting of the costs of large-scale farming would have to include the energy requirements resulting from the spatial reorganization of the labor force. For example, since about fifteen percent of the labor force has to commute to work to manufacture agricultural inputs or distribute food, a comparable proportion of the energy cost of transporting commuters might properly be charged to agriculture. In contrast, traditional Chinese wet rice agriculture at its best produces about fifty calories of food for each calorie of human energy expended in farming. The Chinese ratio includes only human laborers and energy input. The United States ratio does not even include human labor, since it is too insignificant. To make these ratios comparable, the Chinese energy costs would have to be limited to the energy consumed in the production of materials which are used as an input in traditional Chinese agriculture. This energy cost is negligible. As a result, a ratio comparable to the United States ratio would be still greater. Chinese energy

FIGURE 1

VALUE OF FOOD PRODUCTION
PER ACRE OF AGRICULTURAL LAND
(1970 Estimate)

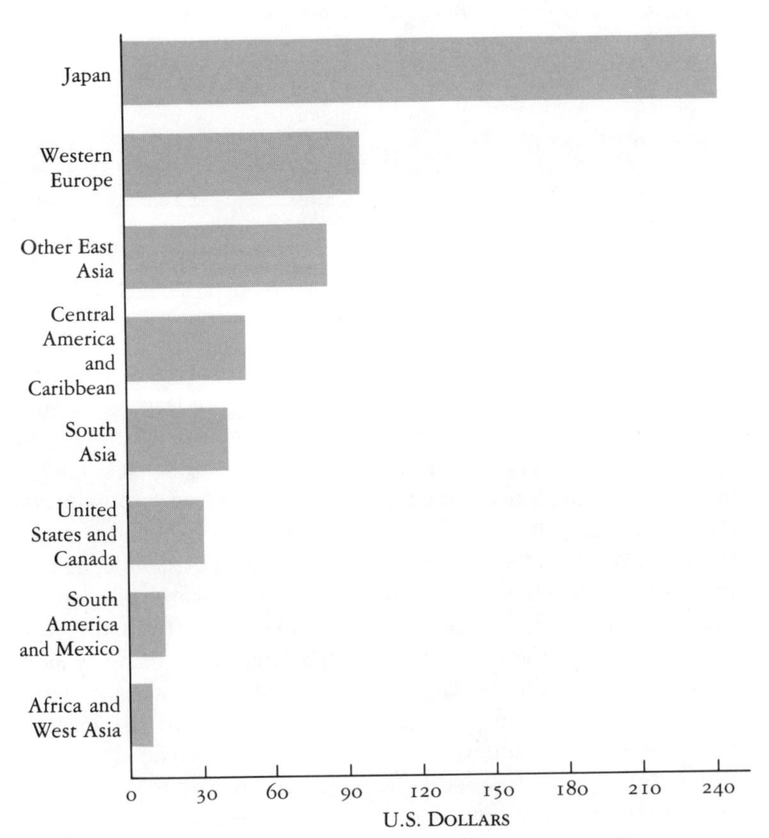

efficiency seems to have been equaled even by primitive food gatherers in the Anatolian Highlands working only with the flint-blade knife. But improving energy efficiency does not mean adopting the life of a hard-working peasant in pre-revolution China any more than it means imitating food gatherers in the ancient Near East. A substantial savings in energy resources could be made very simply by de-emphasizing the wasteful practices of large-scale, profit-oriented farming.

TABLE 1

FARM SIZE AND MARKET DEPENDENCE, 1971

	PERCENT OF TOTAL				
Gross Sales × $1000	Acres Grown	Cash Receipts	Production Expenses	Pesticide Expenditures	Realized Net Income
0 – 2.5	6	2.7	3.7	2	7.0
2.5 – 5	5	3.0	3.2	2	4.6
5 – 10	7	5.9	5.6	5	7.9
10 – 20	14	11.0	10.1	11	14.5
20 – 40	22	20.0	18.4	22	24.6
40+	46	57.3	59.0	57	41.4

SOURCE: USDA Economic Research Service, *Farm Income Situation*, FIS-222, July 1973, p. 71; and Helen T. Blake and Paul A. Andrilenas, "Farmers' Use of Pesticides in 1971," USDA Economic Research Service, Ag. Econ. Report No. 296, November, 1975, p. 6.

Table 1 shows the economic performance of large and small farms. It amply demonstrates the costly nature of large-scale farming in the United States. First, notice that the small farmer earns a substantially larger income per acre while using far less energy-intensive inputs. Since the classification is based on farm sales, the data is seriously distorted in a way which makes large farms appear to be much more efficient than they really are. For example, imagine two farms with identical resources. One is run carelessly and the other meticulously. Because of the higher sales from the latter farm, it must be classed in our data as a larger farm than the inefficient operation. That is, largeness can be an effect and not a cause of the higher efficiency. Because of this system of classification, farms with better soil would also tend to be classified as larger farms. In addition, large-scale farm operations tend to get more for selling the farm product; that is, they have marketing advantages.

Finally, they are more likely to do some of the processing of the crop on the farm.

Banks also restrict the economic performance of small-scale farmers, because they frequently refuse to lend money to small farmers for lucrative crops such as lettuce or potatoes. Finally, the small farm is likely to consume more of its own produce on the premises. Naturally, sales will suffer. The column for purchased inputs also minimizes the difference between large- and small-scale farms. Small farmers pay significantly more for the same input. As a result, the monetary value of inputs on large and small farms represents very different physical quantities. In addition, some of the materials classified as purchased inputs are not used for farm production at all. For example, the United States Department of Agriculture tells us that in their study of large- and small-scale farming, the amount of fuel per acre was about equal. Now, on closer inspection most of the fuel on small-scale farms is gasoline. Apparently many small farmers are writing off trips to town in the pickup as a farm expense when they went to visit friends or go shopping.

Some economists argue that the ability of the small farmer to produce more income with fewer inputs results from the excessive working time put in on the small farm. In reality, many of the small farms are run by part-time farmers with a full-time job outside of agriculture. For the agricultural sector as a whole, more than half of the total farm income reported in 1970 came from employment away from the farm. For small farms the percentage is significantly higher. In calculating the labor cost in small-scale farming, economists do not consider all labor equally. An industry which uses workers who have no alternative possibility for employment is presumed to be less costly to society than one which uses highly skilled trained workers. Small farms use very little high-quality labor measured by market criteria. For example, in one Department of Agriculture study of small farmers in the Ozarks, about twenty percent of the household heads were found to be totally or partially disabled. In short, our estimates seriously overestimate the labor costs of small-scale farming.

In another sense, the labor costs of large-scale farming are seriously underestimated as well. In spite of the notorious underreporting of occupational diseases and injuries, agriculture is still the third most dangerous occupation in the United States—far more dangerous than the work of the policemen who gain our admiration every night on television. In California, more than one out of every five farmworkers employed by giant farming corporations succumbed to an occupational disease or injury in 1971. In addition, many of these diseases and injuries are not ever reported in agriculture. This rate is still about three-and-a-half times the incidence rate in large-scale industry where reporting is much more comprehensive. In fact, during the first few years of the Second World War more Americans died on farms than on the battlefield.

Finally, the quality of food is suffering from the effects of energy-intensive agricultural technology. Products have been indiscriminately bred to increase yield without any consideration of nutritional consequences. For example, teosinte, widely believed to be an ancestor of corn, had a protein content of 19 to 23 percent. In addition, many of these so-called primitive strains of corn had a much higher quality of protein. Lysine is still abundant in native corns in Central America. But with hybrid corn, the lysine content has gone down seriously. With the introduction of hybrid corn, yields increased but protein content fell from 10 percent to a little less than 9 percent in 1946. In addition, the quality of the protein content fell. And while protein content has risen up to about 10 percent since that time, the additional level of recovery of the protein content was accompanied by still further decrease in the lysine content of corn—lysine and tryptophan. Protein content of oats and potatoes also seems to have declined as a result of modern agriculture. The situation in wheat is so bad that some wheat is currently being harvested in Montana which is unfit for milling because it does not have enough protein to stick together when you want to make a loaf of bread out of it.

Vegetable quality is also affected by modern technology. Commercial vegetables are bred for the convenience of mechanical harvesting and for the response to fertilizers. The effect of the former cannot be measured, but the latter seems to have involved a distortion in the composition of trace elements. English vegetables contain about three times as much potassium as they did in 1940 while the magnesium content has fallen to one-half the 1940 level. The production of copper is only one-third as great as in 1940, and sodium is only one-sixth as prevalent as it was earlier.

Large-scale, energy-intensive agriculture is so inefficient that it would not last a single day if it were not propped up by massive subsidies, free research, tax benefits, and discount prices. Even so, the corporate farm is hardly an economic success. In a pamphlet published by the Department of Agriculture, we learn that a typical farm corporation in 1970 had total assets of $160,000 business receipts, and showed a small net loss after paying salaries to its officers. In addition, as Peter Gillingham mentioned, large-scale farming is more likely to go bankrupt than a small farm. And who benefits from all this nonsense? Land speculators reap the profits from rising real estate prices. In fact, most of the profits from farming now come from the appreciation of land values. More important, banks and other sources of agricultural credit have an interest in expensive technology which increases the expenses and dependence of farmers. This should be taken quite seriously. When the farmer falters, financial institutions can pick up the pieces. The biggest beneficiary of large-scale, capital-intensive agriculture is, in the final analysis, capitalism as a whole. The forced migration from the farms resulting from the labor-saving technology has provided employers with a constant stream of cheap labor. The once-independent farm family which prided itself on its self-sufficiency has been transformed into a household of obedient consumers.

Tragically, United States farming methods are being accepted around the world. The myth of agricultural sufficiency is being swallowed by nations which can ill afford to repeat the mistakes of the United States. Only one nation stands out as an exception: China. The quarter of the world's population which occupies that mostly barren portion known as China is showing the world that cooperation, social planning, and a desire to serve the people, are the most important prerequisites to agricultural progress.

Good Engineering Produces Appropriate Technology

Roger Garrett

Engineering produces technology of all scales—large, small, and intermediate; whatever is appropriate. Tractors for agriculture are available in horsepower ratings from 5 to 50 to 250. Since our farms vary in size from 5 to 500 to 5,000 acres, there is an appropriate use for each size tractor produced. And, there is a wide range of appropriate sizes of farms. Five acres of kiwi in the right location would produce a good income. So would 500 acres of tomatoes; but 5,000 acres may be needed for range land for cattle or sheep. To some people small is beautiful; but the spectrum of appropriate technology extends from small-scale hand operations to large-scale operations with complete automation. Whatever the enterprise, good engineering produces appropriate technology.

I heard some of you saying, "If good engineering produces appropriate technology, there must be a lot of poor engineers." No doubt there are some, just as there are poor teachers and poor lawyers and poor doctors and poor practitioners of any other profession. But by and large, what many would blame on poor engineering is not that at all. It may be an obsolete technology because times and conditions change. It may be an unfortunate attempt to adapt to one situation a technology developed for another. Or, it may simply be the best solution acceptable under the circumstances.

To understand more fully why the developments of good engineering are sometimes labeled as inappropriate, let us first consider what engineering is. One widely held definition says, "Engineering is the art of organizing men and of directing the forces and materials of nature for the benefit of the human race." According to the last portion of that definition, engineering deals with the problems of people. Typically, people who have problems have only a vague understanding of what those problems are, and the engineer's first task is to analyze the problem. He must know clearly what results are desired and into what present framework the new design must fit. Next, the engineer must determine what resources are available. He will need materials, equipment, energy, labor, knowledge, skills, time, and acceptance. The list of resources needed points out the fact that his design must function within numerous constraints. The engineer, of course, must have technological capability; and, because many of the required resources involve investment of capital, he must produce a design which is financially effective. Finally, his design must be socially acceptable. This fact is becoming increasingly apparent. Concerns for the engineer's use of limited resources and the impact of his designs on the environment and on our social structures are being made known through legal and political constraints and through the ethical concerns of the engineer himself.

The engineering process produces solutions for specific problem situations. When one tries to apply technology designed for one situation to another situation, problems are likely to occur. A tractor designed for California is not necessarily appropriate for India or Mexico or for that matter even for Missouri. For any tractor there must be a competent operator; service, maintenance, and repair facilities; a supply of repair parts; and a class of tasks suitable for the tractor in sufficient quantity and at suitable intervals. For the California tractor, India may lack a competent operator; Mexico may lack a supply of repair parts; and Missouri may lack a sufficient quantity of suitable tasks for the tractor. In all these cases it would be inappropriate to apply the technology which was designed for conditions in California.

California growers have long recognized the problems of adapting technology designed for other regions. Farm machinery produced by major manufacturers in the midwestern United States is frequently unsuitable for California conditions. California growers have become so adept at modifying that equipment to suit their special needs that they often produce designs for their own. (One of the major designs of tomato harvesters was developed by a grower to suit his needs.) But some people are not so adept at modifying and adapting. All too frequently, equipment sits by the side of the field and rusts away because it was not appropriate for the conditions to which it was applied. If this situation occurs in the United States for a piece of equipment developed in this country, it is not surprising that the same situation occurs when equipment is transplanted to another country and culture.

In the United States, and particularly in California, agriculture has adopted, for some commodities, methods of production which make intensive use of capital and energy. Do these technologies represent poor engineering? There are some who believe they do. However, a look at the history during which these technologies were adopted shows that major steps toward mechanization took place during periods of inadequate labor supply. Furthermore, many developments in mechanization have taken place along with the expansion of agriculture into unpopulated areas, where again the labor supply was inadequate. An adequate labor supply must include a sufficient number of workers, with the necessary capabilities for farm work, available at the proper time, and willing to work for a wage which is compatible with the overall financial effectiveness of the operation. Because this country was rich in resources during those periods of agricultural development, it was easy for people to produce more than they needed. This excess production led to an accumulation of capital. With the availability of capital, along with the availability of low-cost energy, it was natural and proper for engineers to use capital and energy, in the absence of adequate labor, to meet the demands of people for more food at less cost, more variety in their diets, less rigorous labor, better working conditions, more leisure time, and so on. Engineers responding to those needs developed mechanized agricultural techniques that were physically efficient, financially effective and, at the time, socially acceptable. It was good engineering.

But times change, constraints change, resource availability changes, costs and revenues change, social standards of acceptability change, and problems change. The kinds of solutions engineers produce must change too, and the technology of agriculture is changing constantly.

With decreasing availability of energy and increasing unemployment, there is increasing social pressure to allow the unemployed to replace energy and capital in food production. At the same time, burgeoning populations demand more food production, and farm workers demand a higher quality of life. No one—especially agricultural engineers—would deny that most farm workers deserve a higher quality of life, however one defines that characteristic. And no one would deny that all people need food to eat. But there is considerable difference of opinion as to whether using labor to replace energy and machines can achieve a better quality of life and increase, or even maintain, present levels of food production. I'd like to cite three examples. First, consider the case of rice production.

California produced about 3 tons per acre with a labor input of about 2 man-hours per ton. India, on farms without mechanization, uses more than 850 man-hours per ton to produce one-third of a ton per acre. Obviously California's technology for growing rice would be inappropriate for small, terraced paddies. At the same time it would be inappropriate to use 848 additional man-hours per ton in California, if by doing so we can achieve only 11 percent of our present yields. The reduction in productivity cannot possibly increase the standard of living of the farm workers or anyone else.

Now, let's consider the case of the workers who harvest lettuce. When you buy a head of lettuce in the grocery store, about two cents of what you spend goes to the workers in the crew that harvested it. If you are one of the truck drivers or dock hands or office workers or produce clerks or grocery checkers who receive wages for moving that lettuce head from the field to the grocery cart, you receive for your efforts an average wage of about $12,000 per year for 2,000 hours of work, or $6 per hour. To earn a living at that same rate, lettuce field workers must harvest and pack 300 heads per hour. Many of them do. Some harvest and pack at the rate of 450 heads per hour. But it's hard work with a lot of stooping, and they don't always work eight hours a day, or five days a week. And certainly they can't work all year long harvesting lettuce unless they are willing to migrate to different production areas.

Steady employment, with a reasonable annual wage, work that is not unduly arduous, the opportunity to settle down and become part of a community—these are aspects of the quality of life which are difficult for a lettuce field worker to achieve. If he can find work with other crops, if he has the skills to participate in land preparation, planting, cultivating, irrigating, or other jobs, he may be able to stretch his period of employment enough to allow him to stay in one place and accumulate a reasonable annual wage.

You could arbitrarily decide to give him more money for what he does; but, unless you are willing to spend a larger proportion of what you earn for the lettuce you buy, foregoing some of the other things you want, you will demand more wages for your services, too, and the farm worker, in the end, will be no better off. Only by increasing his productivity can he achieve any real and lasting benefit. He could work harder, but it's difficult to see how. Anyway, one aspect of a good quality of life is not having to work too hard. We could alter the way lettuce grows to make his work more efficient. We are certainly trying, but the probable benefits are small.

The only way to make large changes in the worker's productivity is by assisting his efforts with capital and energy in the form of mechanization.

As an engineer, I have tried to develop technology that would improve the lettuce field worker's productivity and allow agriculture to continue to meet the demands of people like you for the lettuce which you have come to expect as part of the quality of life you enjoy. Now obviously, if I increase a worker's productivity and the same total work is to be done, fewer workers are needed. But let's look at a third example to see what frequently happens.

In 1960, crews of mostly Mexican braceros harvested 2.24 million tons of processing tomatoes in California at the rate of 0.22 tons per man-hour, or about 10 million man-hours. Using mechanical harvesters, 7.42 million tons were harvested in 1975 at a rate of about one ton per man-hour or 7.42 million man-hours. Because of political constraints, the workers in 1975 were principally women, students, and green-card-carrying Mexican nationals. The net result, after the introduction of the mechanical harvester, was a threefold increase in production, a 25 percent decrease in labor required, and an increase of 4.5 times in productivity for the workers. You may argue that the social impact of changing to a labor force including women and students is good or bad. But whichever side you take, remember that the change in the labor force resulted from a politically imposed constraint.

I have tried to express and illustrate several important points. Engineers respond to the problems of people and use available resources to solve these problems. The solutions are specific to the problem and to the resources and constraints; caution must be used in attempting to adapt technology to other problem situations. The technology of agricultural mechanization which engineers have developed in the United States contributes greatly to the efficiency and productivity of its agriculture and to the quality of life of its people. Those solutions were and still are appropriate. And good engineering will continue to produce appropriate technology.

Discussion

AUDIENCE PARTICIPANT: Do you have any specific recommendations from all this studying you've done, and is there anything I can do personally to try to implement all these schemes that you propose?

PETER GILLINGHAM: We are not a long-established organization. This paper was an effort done under great time pressure by a very small organization with virtually no paid staff to pull together some of the sources, some of the ideas. Even at that we found that some extraordinary work was going on around here. The Institute for Man and Science up in Covelo, Ecology Action of the Mid-Peninsula with their research mini-farm on the Syntex land in Palo Alto, the Community Environmental Council, the CEC down in Santa Barbara. I happen to have named four groups all working on the biodynamic French intensive method, and there are other methods.

There are many answers to your question. I would say that the jobs are very hard to get because the configuration of institutions now is still in the wrong directions, substantially but not completely, for the things you want to do, the values you want to serve and pursue in your own life. I suppose for a lot of people, to be brutal, I'd say to go out and support yourself some other way while you go to work with these outfits and help them in their work and learn what is really possible. It can give you a whole new dimension. You realize what can actually be done with *x* amount of water on *y* amount of land, and to see these yields produce after the soil has been renewed to health. I think that's the first thing you see, the work, and the job flows from it. Support yourself in any way you can, because you may not be able to draw your financial sustenance from agriculture as such. This is perhaps not a practical recommendation for some people, but it seems to work with the people we know. A lot of people are doing it.

AUDIENCE PARTICIPANT: I'd like you to comment on the number of people and the number of hours involved to produce the same amount of food from a French intensive or other method in contrast to agribusiness.

PETER GILLINGHAM: We already have the beginnings of structure in which a few people are on a farm and a substantial number of people from urban and suburban areas go out and help at peak-load times with planting and harvesting, and so on. The Soil Association in England actually runs an operation that mobilizes about six thousand people a year. It places them on organic farms. Similarly, one of the things that I never thought about until Schumacher said it, because I'm ignorant in this field, is the idea that to conduct only agriculture in rural areas is economic idiocy as well as social idiocy. All too often we only consider agriculture and we don't think of the needs of other forms of economically productive activity and the appropriate kind of technology.

MICHAEL PERELMAN (addressing Roger Garrett): Dr. Garrett's lettuce example, when he talks about the yield of lettuce; shouldn't we consider the fact that, with our profit-oriented marketing system, one-third of the lettuce is thrown away on the way to the supermarket. Also, when we talk about the labor going into the lettuce, we do get advantages from growing lettuce all in one place. But these advantages are balanced, at least in counterpart by the need for the middleman you talked about. It's true that the labor going into lettuce production is extremely small, but there is a prerequisite of other types of labor which is going in there. Finally, in talking about Indian lettuce yields, are you taking into account that these lettuce fields are growing other crops the rest of the year? That is, that they are not just growing lettuce, but double-cropping, or using a relay cropping system.

ROGER GARRETT: There seems to have been a misunderstanding. First of all, I did not talk about the yields of lettuce anywhere, certainly not in India, because as far as I know, India does not grow the type of crisp head lettuce that we take for granted in this country. In fact, very few countries do grow our type crisp head lettuce. I did talk about the labor inputs in lettuce and how much of the cost of a head of lettuce actually goes to the guy who's doing the harvesting in the field.

Your comment about lettuce growing in one place as opposed to, I suppose, growing it where the population centers are, deserves some comment. We do grow some lettuce in the fringe areas around the population centers as much as is possible there, but if we depended on those, then the people in those population centers would have lettuce available to them only a few months out of the year when the climatic conditions were proper for that growth. It's only in California's ideal climate, and in the western lettuce states which include Arizona, New Mexico, and Texas, but primarily California and Arizona, that we have the climatic conditions such as to allow the growth of lettuce on a year-round basis, so that you can go to the supermarket anywhere in the United States, any time of the year, and get a head of lettuce.

VASHEK CERVINKA (addressing Michael Perelman): Dr. Perelman, you indicated in your comments that the value of crops on an acre of land is higher in the United States and Canada than in Western Europe or Germany for example. Would you explain why, please?

MICHAEL PERELMAN: I was talking about total land base. Europe and Japan do not have, in the first place, an excessive livestock system as we do, which reduces the value per acre. Also, they have better social planning so their cities don't sprawl all over the area, and farms are not knocked out of operation. They devote a much, much larger percentage of their total land to growing crops. For that reason, to make the statistics comparable, I used figures on output per acre rather than output per unit of land base. And here we find out that European countries do substantially better in general on their crops than do United States farmers. Including, mind you, farmers in Central Europe, farming under what some people would consider a handicap, with which I would disagree, of a Communist form of agriculture. I think the low yields of the United States represent a concentration on profit rather than production. I think they represent an insensitivity to the needs of people for good-quality food, and I think they represent, in effect, a tragic confirmation of the total bankruptcy of the agricultural system, as well as the total economic system that we have today.

VASHEK CERVINKA: I feel I have to comment on that response. I believe that the reason why farmers in Europe or Japan are achieving higher yields per acre is because their agriculture is much more energy-intensive than agriculture in the United States. They have more tractors. They are using more fertilizers. I think energy-intensive agriculture in Japan and Europe is the reason why they are achieving higher volumes or higher yields per acre. I explained my views of agricultural development in my presentation and there are certain facts which I think we have to deal with here. Again I have to say certain things which are not popular here. I strongly believe we cannot do without energy in agriculture. One reason is that without energy inputs we cannot achieve higher yields. A problem is that we have to feed a continuously growing population and thus we have to increase production. It means we cannot produce high yields without giving inputs into the production.

This leads me to another problem I want to discuss. There is so often discussion about organic farming or farming with fertilizers. Some people may disagree with me, but I do not believe we can do without fertilizers in agriculture. There was mentioned today a study which was done in the Midwest which indicated that organic farming is achieving higher yields than standard agriculture based on fertilizers. I am quite familiar in detail with that study. I would like to give a few points about that study. In that study there were conventional farms, and organic farms, sixteen pairs. And they were growing four crops. They were growing corn, oats, barley, and soybeans. In three crops, in corn, oats, and barley, production per acre of land was over in organic farms. The yield was either the same or slightly higher on organic farms. But I had one problem with that study because I saw the data was misleading. The fact is you can grow soybeans without any nitrogen fertilizers because the soybeans by nature are fixing the nitrogen. So I think this comparison of organic farming and fertilizing farming wasn't really fair. It doesn't mean that I feel we should do without organic farming. I am strongly convinced that we cannot do without fertilizers.

AUDIENCE PARTICIPANT: Dr. Garrett, you mentioned that agricultural engineers address themselves to problems of people. I would like to venture to say that in the future and starting very soon, the scientist in general, and the agricultural engineer in particular, has to address himself to the problems of all the people, rather than to the problems of certain people involved in farming and profit-making. I think the scientist in general has to take the role or part of the role as leader and look at the world with a more holistic view, rather than just one of profit or production orientation.

ROGER GARRETT: It may surprise you and a lot of other people to realize that we are working on the problems of people other than simply the farmers themselves. We're working on the production of food that we all eat. Since the largest portion of our population exists in urban areas rather than on farms, it's for those people that we are working when we try to help farmers to provide food that they want and need.

AUDIENCE PARTICIPANT: It seems that one of our most serious problems with our existing technological system as it is today is there isn't within it a self-correcting mechanism, a process by which it can examine itself. And I think it is important to keep in mind the element of self-criticism. What is your response, Dr. Garrett?

ROGER GARRETT: I certainly agree with you that we have a responsibility as engineers to subject ourselves to self-criticism; that's why I am here today. We definitely recognize a responsibility for what we do and we recognize that what we do has a tremendous impact on a lot of people. We recognize that there is also an impact from doing nothing. And that impact, in our judgment, would be worse than doing what we are doing. We have to make that kind of a judgment and we have to accept the responsibilities to live with those decisions after we have made them. And that's all we can do. We can't know whether what we are doing is right necessarily, and we can't know what will happen if we do nothing unless we choose to do nothing. We have no basis by which we can compare the results of our actions. We can only surmise what is going to happen and we have to allow ourselves to be criticized by others, and we have to subject ourselves to self-criticism.

And we do, certainly, try to redefine problems. In many cases we end up solving a problem considerably different than that which was initially proposed to us, because we find people are too close to their own problems to recognize exactly what they are. We tend to look at those problems in a larger context. Problems frequently come to us from farmers, but we view them from the standpoint of all of society, and we may not solve the problem as that particular farmer understood it initially. So we do try to redefine problems to the best of our ability. Our objective in all of this is to create new alternatives which society can use to solve its basic problems of food production and those solutions will be right for some places and wrong for others. We have to hope that people will have the wisdom to know where it is right and where it is wrong.

AUDIENCE PARTICIPANT: I want to address my question to Phil LeVeen. I'm interested in what practical alternatives you see farmers taking to fight this treadmill effect which I thought you described well.

PHILLIP LEVEEN: I wish I had the answers. I think it's easier to describe the problem than it is to give a solution toward it. Clearly, in the industrial economy, especially in the more concentrated sectors of our economy, many of our firms have figured out a way of controlling the rate of introduction of new technology. Farmers are much more vulnerable because of their competitive situation and their isolated situation. To confer monopoly powers on farmers to make decisions like Ford and General Motors can make, cannot be an ideal solution.

I'm particularly concerned that the problems that we are dealing with are not simply agricultural but economy-wide. It may well make sense for the Ford Motor or other automobile companies to give up some of their use of energy so more can be made available so farmers don't have to stoop in the fields. I think we have to be careful in talking about the whole system you know, and not just pick out one part of it. We must make sure that we keep the whole thing in mind. As far as practical solutions, a lot of the problem we are dealing with is concerned with institutional changes that would involve some of our most basic principles. Problems of granting and conferring monopoly power on farmers, for example, violates most of our antitrust. We do this in some sense in our marketing orders and farmers will deny this, but in fact marketing orders function in the way I've described. Our commodity programs and government programs to some extent have given a farmer a collective voice in the kind of price he gets, but they have done nothing to stop or interfere with the basic process with regard to his adoption of new technology.

I think that the problem is society-wide and when I talk about social justice and equity I was thinking far more in terms of a general problem than I was in terms of a specific resolution within the farm sector or within the marketing sector within the farm sector. I was thinking about general processes, creating general processes where all individuals that are interested can get involved and help create their own answers. I don't think there are answers. I think these answers come out from a different kind of process than we now have.

AUDIENCE PARTICIPANT: Dr. LeVeen, you have made a traditional economic argument when you are talking about price competitive and cost competitive situations, especially in the Midwest. How are these costs calculated? What would happen to our society, to our economy, if the true total cost of doing things with fertilizers, insecticides, you know, if we could institute a mechanism for resource depletion and the cost of that?

PHILLIP LEVEEN: I meant to make that point and I'm glad you gave me a chance. There's no doubt that what we've seen is a response in part to relative prices. Labor in the Midwest in particular was expensive and land and capital cheap. Artificially cheap in the sense of long-run supply of these resources. The environmental considerations impose no specific cost on the farmers, so in general you could argue the farmer's working on the basis of a set of costs which are not relevant, that his own private calculations are different from what the society bears. That's the standard economic response to what you are saying. No doubt the energy crisis deepens and the prices of capital and fertilizers and pesticides grow relative to the prices of labor and land. The kinds of technology that will be used, and in particular in the family farm states, will change. I think they will change to some extent too in California, especially those that do not threaten the basic relationship between capital and labor. For example, we now have in California an increasing interest in integrated pest controls. These pest programs are being developed at the University of California at Davis and

Berkeley, and to some extent they have been very useful in reducing farmer dependence on certain kinds of chemicals and certain kinds of crops. Now, the interesting thing is that these systems tend to be beneficial to large farms rather than to small farms. The delivery system has been designed, perhaps inevitably, in a way that it's supportive of the larger farmer rather than the small farm. And so getting away from energy use in agriculture through reducing our dependence on fertilizers and pesticides may be possible. But it doesn't necessarily mean it's going to support small farming.

AUDIENCE PARTICIPANT: In one area in California we are working on a reclaimed water project. But reclaiming water may be very energy-intensive, as well as toxic to the soil and dangerous to health. What may seem to be an ecological answer to a problem may instead be a trap. How would you respond? How, after you've drawn these issues of what you do about preserving small farms, diminishing resources, mainly water, an alternative form of technology to supply that water, would you pull all that together? Do you have any suggestions?

MICHAEL PERELMAN: You know what they say about fools rushing in! Now our problem here in agriculture is that we have a very fragmented system and it's a problem that is endemic to the market system as a whole. That is, I suspect we started out a long time ago and talked about how you keep these nutrients, resources, water, moving more effectively and we have put city engineers together with farmers and agronomists so they would come up with an idea of using reclaimed water. Now what happens typically is that the market gets us into a bind and then all of a sudden people like you are called upon to make a decision right now. And that decision almost becomes impossible to make because it's a decision which is affecting the lives of many, many people. People who are going to eat the food, people who are going to be affected as far as farming, water supply, and the like. Now the only way that I can see that you could really tackle a question like that is if we start talking as a group, as a society. Here we have a basic resource, we have a certain pool of skills, we have a certain type of life style which we want. How are we going to do it? Now those questions will not be asked by a small group of people who are the board of directors of this company, or the bureaucrats of that company.

What we need is a total system of social planning so we can ask these questions in advance and then we have the information that we can give you so there isn't a big crisis question thrust upon us. The problem is that questions like this are bound to arise and they are going to arise again, and again, and again, one after another until we really grasp this problem at the roots and say what it is we want to do with our lives, and we take control of our lives and we start doing it.

AUDIENCE PARTICIPANT: I have a question for Phil LeVeen. I wanted to ask you about the tremendous price differentials between organically grown things, especially in terms of grains and rice even within community food systems in the San Francisco Bay Area. I noticed sometimes almost a 100 percent mark-up, and if that cost isn't ending up in the pocket of the organic farmer, I'm wondering where it is going?

PHILLIP LEVEEN: I've looked at the problem a little bit and you're right. There's a tremendous premium, there's a sum premium for paying for organic food. Whether it comes in fact from an inefficient marketing system or whether organic farmers do need more than traditional farmers, I can't say. I do believe that organic agriculture probably, at the farm level, is not more expensive than traditional farming. There are problems with organic agriculture though, when you put it into a distribution system where there is rot and there is no preservative in the organically grown things, so they rot more. The system they are employing may not be as effective as Safeway's tremendous network in getting rice and packaged things to you. Also, it's a small thing and there are so few people in it that there may well be a monopoly in the thing, in the system itself. I don't know. It's such a small system that it's very difficult to figure out what is going on.

ROGER GARRETT: I'd like to comment also on that if I could. One of the advantages that a lot of people see in organic foods is a difference in flavor, a difference in texture, better quality you see. One of the reasons why this occurs very frequently is because you are very close to the point of supply. The food is harvested closer to the peak of perfection. You get it that way. But at the same time you are dealing with a commodity that is highly perishable. Now one of the things that our high-technology agriculture has done is make it possible for us to get food to masses of population that are considerable distances from the points of supply. In doing so we have frequently had to resort to harvesting tomatoes in what we call the mature green condition. It matures on the way to market or holding in storage. Now, you have to contrast that sort of technology with no tomatoes at all. Don't compare it to little tomatoes grown in your little hothouse window box. If you can do that fine, go ahead and do it. But in New York City you'll find that you can't grow tomatoes that way all year long. And you can't grow them on Long Island all year long either. So if you're going to have tomatoes all year round, and if you want that as part of your quality of life, then those tomatoes are going to be a litle hard sometimes, but they're going to be there and you can have that or nothing. That's the situation we have right now.

AUDIENCE PARTICIPANT: Dr. LeVeen, what would be the first priorities for alternatives in state and local policy so that the political, economic climate of California agriculture could accommodate intermediate technology?

PHILLIP LEVEEN: I don't think you can answer questions about California agricultural-political economy problems. We have to look at this holistically and the only way we can look at it holistically would be to do something in political economy that would be comparable to Jimmy Carter's budgetary plans. He says, "Start out every year with a clean slate and see what you can come up with." So I would say in political economy we have to do the same. We're going to have to start out with a clean slate. Just wipe the whole thing clean and then build it up and see, build it up with the whole purpose of asking how each component serves the people as a whole. That would be my response.

AUDIENCE PARTICIPANT: I would like to ask all you gentlemen, why in discussing alternative approaches to producing food you have not considered alternatives to consuming food? I would like to suggest that the solution is not to have a lot of extra for everyone but rather for everyone only to take enough. And I would like to ask if any of you experts have any figures on the amount of wasted food there is, and I include meat because I consider meat a wasted food.

PHILLIP LEVEEN: There was a study done in Tucson, Arizona, I think it was, in which they found that a substantial amount of food was wasted. Plus, if you look at the distribution system, the figure was one-third which I used for lettuce, that is the outer leaves that are discarded as a matter of course. That doesn't include the amount that is thrown away as they age and go bad during the marketing trip. You're right on the amount of waste.

VASHEK CERVINKA: I will answer your question. I came to this country about eight or nine years ago and my first impression, and this was an experience shared by other people who migrated to this country, was of fantastic wastage. There's waste of food, of gasoline for unnecessary trips by car, waste of packing material, newspapers, Sunday newspapers, it's fantastic waste. That's the thing. Now I agree there's waste. It may be illogical, for example, to transport lettuce to New York, but now there are two questions. Either consumers agree or accept that they will be consuming fresh vegetables just for a few months a year, or they will have a supply for twelve months as it is now. That is the question.

We agree that there is a fantastic waste of energy. What fantastic consumption of energy in the processing of food! But the question is, is it the fault of agriculture that food is processed or is it a pattern of our life? I don't know really how technically feasible and economically feasible it would be to supply fourteen or fifteen million people in New York with fresh vegetables. How it would be to supply Chicago or even Los Angeles or San Francisco just with fresh produce, not processed produce; I really don't know. Going to organic farming, I think it has very much something to do with the scale of operations. If you have your own garden, if you are just growing organic food, you may be quite efficient. But take the farm. You may be using just organic matter but now you have to transport this organic matter to apply to the field from a certain distance. And this distance may be quite large. You are then talking about organic farming, but actually you are using energy for transporting organic material to the farm. The question is, should we use it for the transportation or should we use it for fertilizers? It is a question of economic binds.

AUDIENCE PARTICIPANT: I have a question for Roger Garrett. Are agricultural engineers giving adequate self-questioning to their assumptions on what the problems are in agriculture? We must avoid the problem that Phil LeVeen mentions of designing a system which cultivates individual rationality but collective absurdity. Is this the weakest link in engineering?

ROGER GARRETT: I think there are a lot of people who have the misconception that an engineer does nothing but build machinery. We do some of that certainly, experimental equipment, but we have a lot of other things. Energy use, conservation, production, are high on our list of priorities at the present time. We have projects involved in creating gas materials, gasification we call it, from waste materials that are left over, anaerobic digestion of animal waste to try to produce our own energy. I think that some day in the future we are going to see agriculture totally energy self-sufficient. We consider that a high-priority need for engineering. There are a lot of other things; I could go on, but I won't. We do try to ask ourselves what are the most important things and frankly that's why we haven't spent a great deal of time on small farms. Those are individual problems that need individual attention and we don't have the resources to devote to the problems of a very small group.

AUDIENCE PARTICIPANT: I have one further related question. Do you have difficulty getting government grants for things that, from an engineering standpoint, you think are really worth pursuing but happen not to be stylish?

ROGER GARRETT: That's a perennial problem. Yes. We are funded by state government, by federal government, and by grants from industry. Unfortunately, a large portion of our research activities is funded by grants from commodity groups that are interested in problems of their own. We try our best to avoid getting ourselves into a position of doing just what someone else wants us to do, but we do have a lack of funds to do the things we would happen to consider most important to do.

IV

THE ETHICAL BASIS OF INTERMEDIATE TECHNOLOGY: SOCIAL IMPLICATIONS

John Coleman, S.J.
Isao Fujimoto
E. F. Schumacher

INTRODUCTION: *Yvonne L. Hunter*
MODERATOR: *Desmond Jolly*

Introduction

Yvonne L. Hunter

When reading almost any passage from *Small Is Beautiful,* or listening to Dr. Schumacher speak, one is struck that his thoughts are founded upon a strong ethical and moral base. Upon more careful probing, one also feels that a deep "spiritual" force is also included in the ideas. The former is easy to identify; the latter more elusive.

For example, in his main address, Schumacher speaks of ethics and morality when he says:

> What can we do to maintain and to increase the range of personal freedom? In England, it used to be said that people were free to dine at the Ritz or to sleep under the bridges of the Thames. This is the kind of philosophy that enraged Karl Marx and many others. You don't have to be a Marxist to get angry at that. Now, why are these people who are "free" to sleep under the bridges of the Thames actually forced to do so?

In general, intermediate technology invokes ethics in its call to action to improve the quality of life, to protect the environment, and to ensure that we help, not harm, developing nations. In fact, the ethical basis of intermediate technology is very similar to the ethics, morality, and altruism that surrounds the environmental movement.

The spiritual basis of intermediate technology, on the other hand, goes much deeper. It is spiritual in that it goes beyond the materialistic. In one sense, Schumacher's ideas imply a spiritual re-capturing of old ideas—self-reliance, working with the soil and one's hands, all of which may lead to a more fulfilling and ennobling life. In another sense, Schumacher's ideas go still deeper. In the final paragraph of *Small Is Beautiful,* Schumacher suggests an answer to the often-asked question, "What can I actually *do?*" He replies:

> The answer is as simple as it is disconcerting: we can each of us, work to put our own inner house in order. The guidance we need for this work cannot be found in science or technology, the value of which utterly depends on the ends they serve; but it can still be found in the traditional wisdom of mankind.*

It is this continual reference to the wisdom of mankind and inner truth that adds an extra dimension to Schumacher's ideas. A recent convert to Catholicism, Schumacher frequently quotes Catholic theologians in his works. He is, according to John Coleman, who shared the Ethics panel with him, a contemporary voice of Social Catholicism. Coleman describes Social Catholics as "conservative radicals" or "reactionary radicals" who mix the old with the new, with all the risks involved. Chuck Fager, writing in the March 13, 1977, issue of the *San Francisco Bay Guardian,* discusses Schumacher, his religion, and his message at length. In a very provocative article, Fager probes the spiritual side of Schumacher.†

Being a Catholic, though, is only one dimension of Schumacher's spirituality. He is a student of Buddhism and has studied Gandhi's philosophy extensively. His chapter called "Buddhist Economics" is now a classic, and its ideas permeate his other works. In truth, the ethical basis of intermediate technology and E. F. Schumacher is a blend of east and west, old and new, Catholicism and Buddhism, radical and conservative, with a good dose of mysticism and metaphysics added for good measure.

The three speakers on the Ethics panel each take a different look at the ethical basis of intermediate technology. John Coleman gives a rich and most insightful look at Schumacher, his ideas, and their relationship to ethics, social justice, equality, and morality. Coleman calls Schumacher's a "wise and humane program of a man who knows much about both virtue and wisdom, two qualities in this

---

*E. F. Schumacher, *Small Is Beautiful: Economics as If People Mattered.* New York: Harper & Row, Perennial Library, 1975, p. 197.

†Chuck Fager, "The Genesis of Small Is Beautiful," *San Francisco Bay Guardian.* March 13, 1977.

world which are very much subject to the classic economic problem of scarcity."

Isao Fujimoto speaks to the relationship between technology and ethics. He discusses the quality of life, maintaining that the type of technology we have influences the kinds of values with which we live. Conversely, he says that the types of ethics by which we live can also influence the kinds of technology we have.

Finally, Dr. Schumacher speaks to the need to address the problems of remote and impersonal institutions and the need to promote better interpersonal relationships. To these ends, he says "small *may* be beautiful."

Moderator Desmond Jolly began the session when he said, "This panel speaks specifically to the question of ethics." He continued:

> Ethics in a technological society is more or less part of the waste of our system. We no longer really involve ethical discussions as a fundamental part of the system. It's more or less an afterthought. This was not always true. If you go back to the older civilizations, the Greek and Egyptian civilizations, for example, you see that they spent a great deal of their time invested in the search for ethical relationships of various elements of the social-economic system to each other. So, it seems to me very fitting that when our system becomes threatened with ecological disaster, with social disruptions, and with personal disintegration, that the question of ethical relationships should once again assume significant proportions.

Ethics and Intermediate Technology

John Coleman, S.J.

I see Schumacher's *Small Is Beautiful* as both a book on development economics and on social ethics. Development economics raises questions to ethics and vice versa. When the book appeared in 1973, there was already widespread questioning of the assumptions of the models of development economics which dominated the first United Nations decade of development in the 1960's. In the spring of 1972, with the publication of the Club of Rome's report on *The Limits to Growth,* economists were signaling the need to attend to the finite energy resources of the planet, especially the fossil fuels. Alternative energy supplies such as nuclear energy carry extraordinary risks of leakage of waste materials. Moreover, the specter of thermal pollution hangs over any future premised on unlimited growth in productivity.

The energy crisis raises anew the classic economic problem of scarcity and the correlative ethical question of allocation of scarce resources according to some canon of distributive justice, such that allocation is not determined by already existing wealth or power. In no moral system does might or mere prior possession make right. A finite world of energy forces us to examine our principles of distributive justice. Either the rich and powerful cut back on energy use in some real redistribution of world wealth or the poor and relatively powerless trim their expectations. The predictable energy resources cannot support global industrialization on the model of Western Europe, Japan, North America, and the Soviet bloc. It is not even clear whether they can support the continued maintenance of these developed systems at their present levels.

Arguments for intermediate technology based solely on energy scarcity and applied uniquely to the less-developed world could, however, skirt entirely the question of distributive justice. By what title in justice do the developed nations consume the amount or percentage of energy they presently use? By what title do they use energy sources located in the developing nations? What moral claims do less-developed countries have on a greater percentage of world energy resources, necessitating a cutback by the more powerful countries? Does the adoption of intermediate, small-scale technology by the less-developed nations entail their permanent relegation to second-class world citizens cut off from a material base for psychic and social mobility and growth? Can they be persuaded to lower expectations in the face of unabated affluence in the developed world, communicated to them through a worldwide media network? By what moral claim can they be asked to lower expectations if the developed nations do not follow suit in programs of what Ivan Illich has called "convivial austerity"?* The first ethical basis for intermediate technology is that it be applied both to developed and less developed countries.

There is, unfortunately, not much agreement in current debates about principles of distributive justice as applied to a world economic order. Schumacher, for example, never uses the term "distributive justice," except in evocative fashion. He seems to be relying on some unclarified norm of equality when he calls for tools cheap enough to be accessible to everyone and for production by the masses instead of mass production. He presents no priority rules for the situations where varying moral claims come in conflict. Intuitive notions of distributive justice are notoriously flabby. I will return to this point later in my remarks.

A second assumption of the development model of the 1960's which has come under attack, is the priority given to urban-based industrialization over agriculture. This has led in Third World countries to a dual economy: a center of urban elites, largely dependent on western-style education, technology, and expertise; and the marginalized rural poor. Such dual economies, far from being development, represent a kind of neo-colonial dependency. Dual economies have resulted in mass migration to the shanty towns which ring the Lusakas, Rios, and Cairos of the world, and have

*Ivan Illich, *Tools for Conviviality,* New York: Harper & Row, 1973.

resulted in mass unemployment. Once again, the question of distributive justice is raised, this time internally for the developing nation.

We have become aware by the examples of the so-called "economic miracles" of Brazil, and to some extent Korea and Mexico, that growth in GNP or per capita income may be unrelated to any real internal redistribution of wealth. As Dudley Seers has put it:

> The questions to ask about a country's development are therefore: what has been happening to poverty? What has been happening to unemployment? What has been happening to inequality? If all three of these have declined from high levels, then beyond doubt this has been a period of development for the country concerned. If one or two of these central problems have been growing worse, especially if all three have, it would be strange to call the result "development" even if per capita income doubled.*

In an important address to the World Bank meeting in Nairobi in 1973, Robert McNamara spoke of the need to shift development priorities from urban industrialization to agriculture, with special emphasis on increasing the productivity of the small-holder. He saw six necessary social changes as preconditions for this shift in emphasis toward agriculture:

1. Land and tenancy reform. (Note that this involves redistribution of both wealth and power.)
2. Better access to credit by lowering interest rates.
3. Assured availability of water through irrigation.
4. Expanded agricultural research and extension facilities for education which give priority to low-risk, inexpensive technology.
5. Great access to the public services of transportation, education, health care, electrification, etc.
6. Most critical of all: new forms of rural institutions and organizations that will give as much attention to promoting the inherent potential and productivity of the poor as is generally given to protecting the power of the privileged.†

---

*Dudley Seers, "The Meaning of Development," in Charles K. Wilber, ed., *The Political Economy of Development,* New York: Random House, 1973, p. 7.

†Robert S. McNamara, "Address to the Board of Governors," Nairobi, Kenya, September 24, 1973, World Bank Reprint, p. 13.

McNamara's strategy places special stress on the creation of jobs, intermediate technology, and the production of necessities for the majority of the population rather than luxuries for an elite. These values have priority in any conflict of values. (Note also McNamara's sensitivity to the issue of power and its redistribution.)

A final attack on the development assumptions of the 1960's is related to the question of human costs. As asked forcefully by Peter Berger in his book *Pyramids of Sacrifice,* by what title can we demand extraordinarily costly renunciations of fundamental political or civil rights (the case of China) or economic and social rights (the case of Brazil) by this and the next generation in the name of future generations? * Each of the "economic miracles" of the 1960's entailed mass denial of fundamental human rights, either civil or economic rights. Berger pleads, then, for participation by the masses in deciding questions of development policy and the amount of renunciation that will be required of them. He raises implicitly, but never answers, two important ethical questions: Are there some human rights—economic, social, civil, or political—so fundamental that they cannot in justice ever be denied; if so, what are they? And, on what basis is a moral claim made in the name of future generations? It is only by grounding this claim that we can feel and push obligations and responsibilities now for foreseeable catastrophes or costly renunciations which we personally may not live to experience, or demand renunciations now of this generation to alleviate more costly renunciations in the future. Neither question is adequately raised and answered in Schumacher's book.

I will now focus on Schumacher's model of an ethical basis for intermediate technology—what it includes and what it leaves out. I can best do so by ranging a series of three sets of ethical propositions distilled from *Small Is Beautiful.*

1. Schumacher asserts a strong interconnection between personal and social ethics by invoking the classic ethics of virtue in emphasizing temperance, justice, courage, and prudence. He is no utilitarian who believes that private vices can become public virtues or that the additive sum of self-interests can yield a common good. He will not allow a radical separation between means and ends. Thus, he cites with approval Gandhi's strictures against those "who dream

---

*Peter Berger, *Pyramids of Sacrifice: Political Ethics and Social Change,* New York: Basic Books, 1974.

of systems so perfect that human persons would no longer have to be good."

Small Is Beautiful speaks of the need for the purification of character by "finding a center and striving after the good." Schumacher argues for renunciation and sacrifice and attacks the vices of greed and envy. He knows that excessive wealth, like power, corrupts. He seems to argue for an Aristotelian mean between luxurious wealth and miserable penury. He calls for a pursuit of wisdom and ascesis by maintaining that "the cultivation and expansion of needs is the antithesis of wisdom: it is also the antithesis of freedom and peace."

I think Schumacher is right in seeing a strong connection between personal and social ethics. He joins a long list of thinkers, including Plato, Aristotle, Augustine, Aquinas, Montesquieu, and James Madison, who knew that a strong commonwealth rests on a virtuous citizenry. Schumacher, however, does not address forcefully the classical ethical questions about whether virtue can be taught or legislated, or, in any other way, coerced. He leaves unasked the question of the motor for achieving moral regeneration in an ethically lax and unvirtuous society. Do we rely merely on personal conversion and good example to bring in the wise and ungreedy society? Do we try to coerce it by law and imposed discipline? Is a spiritual revolution a precondition for an effective technological revolution? If so, what is its carrier?

2. Schumacher's meta-economics serves simultaneously as a kind of meta-ethics. By appealing to a proper understanding of the human person and his/her place in the natural environment, Schumacher champions health, beauty, permanence, simplicity, nonviolence, and work as an arena for the human need for creativity as paramount values over productivity. While I applaud his contention that economics cannot be divorced from ethics, I wonder if he does justice to values other than productivity which have accompanied the modern industrial process and might be in jeopardy with its loss of scale. I am thinking of the freedoms for personal development; private idiosyncrasy and group pluralism; institutional bases for continuous innovation in the legal, cultural, scientific, and technical realms; empathetic communication across class, ethnic, and national boundaries; and social and psychological mobility.

Social mobility in the advanced industrial nations, at any rate, and redistribution of wealth were tied to large-scale organizations:

labor unions and workers' parties. Large can mean anonymity and a schizophrenic divorce of the public and private realms. It has also entailed, as Peter Berger argues in his book *The Homeless Mind,* the discovery of universal notions of human dignity, "the discovery of the autonomous individual with a dignity deriving from his very being, over and above all and any social identifications."*

The Homeless Mind might serve as the anti-book to Schumacher's *Small Is Beautiful,* although the two are not in simple contradiction. Berger argues persuasively, it seems to me, that there is a distinctive kind of modern moral consciousness logically tied, in what he calls a "package," with large-scale technological production and bureaucratic agencies of administration. Small has often meant parochialism and a closed ethnic or clan morality. Large has brought about an open and universalizing morality, one that seeks to specify and link human rights to bureaucratically identifiable rights to which one can lay claim in law and fair, impersonal, processes. If there are limits to technology and bureaucracy because of their threats of alienation, depersonalization, and meaninglessness, there are no less limits to scaling them down. In Berger's language, we would pay a high moral cost in substituting particularist notions of personal and group *honor* for the more universal concept of human *dignity.* This moral consciousness, as a social achievement, is related to the emergence of modern structures of large scale. To cite Berger:

> The unqualified denunciation of the contemporary constellations of institutions and identities fails to perceive the vast moral achievements made possible by just this constellation. . . . Anyone denouncing the modern world, *tout court,* could pause and question whether he wishes to include in that denunciation the specifically modern discoveries of human dignity and human rights. The conviction that even the weakest members of society have an inherent right to protection and dignity; the proscription of slavery in all of its forms of racial and ethnic oppression; the staggering discovery of the dignity and rights of the child; the new sensitivity to cruelty, from the abhorrence of torture to the codification of the crime of genocide . . . ; the new recognition of individual responsibility for all actions, even those assigned to the individual within specific institutional roles, a recognition that attained the force of law

*Peter Berger, Brigitte Berger, and Hansfried Kellner, *The Homeless Mind: Modernization and Consciousness.* New York: Random House, 1974.

> at Nuremberg—all these and others are moral achievements that would be unthinkable without the peculiar constellation of the modern world. To reject them is unthinkable ethically. By the same token it is not possible to simply trace them to a false anthropology.*

Again, Berger refers to the assumption of maximalization in technological systems—more product for less expenditure. He ties this to a built-in innovative tendency in modern, large-scale technological systems. I wonder if Schumacher has reflected enough on the extent to which the discovery and dissemination of an effective, imaginative, and innovative intermediate technology depends on relatively large-scale institutions and structures of research, teaching, and publication of technical discoveries. Indeed, intermediate technology may be tied parasitically to the same research institutions which have created modern large-scale technologies.

3. Schumacher's third principle of ethics is primarily a principle in social ethics: subsidiary functions. He argues to the convenience, humanity, and manageability of smallness where action can remain a highly personal affair. As he asserts it, "people can be themselves only in small comprehensible groups." I agree with this principle, largely enunciated in Pius XI's encyclical *Quadragesimo Anno.* I would go further than Schumacher in grounding this principle, beyond convenience and efficiency, in a fundamental right to intermediate voluntary associations whose authority is not derived from the state and in the espousal of a pluralism in societal authority.† Associations, as Durkheim saw, act as buffer groups between the state or large corporations and bureaucracies and the individual to provide a sense of meaningful belonging and a carrier of claims for individual rights against the state and large bureaucracy.

What does Schumacher neglect in his treatment of the ethical basis for an intermediate technology? First, he is relatively silent about the relation of power to both individual freedom and social justice. Power is never, in itself, a moral claim to any right. It is, however, the necessary vehicle in an imperfect world to make sure

*Ibid.

†For an ethical justification of a right to voluntary association and pluralism in social authority, see John Courtney Murray, *The Problem of Religious Freedom*. Westminster, Md., 1965; and James Luther Adams, *On Being Human, Religiously: Selected Essays in Religion and Society*, Boston: Beacon Press, 1974, pp. 57–89.

that one's moral claim, justified on other grounds, gets a fair hearing. Reinhold Niebuhr is correct when he sees the creation of a balance of power checking other power centers and the organization and creation of power by the powerless as the instrument of justice in the social order. Given large-scale power groups in nation states, multinational corporations, and large bureaucracies, a balance of power for the powerless may not be possible if we always think on a small scale. Schumacher says little about the ways less developed nations can achieve a collective bargaining position vis-à-vis the developed world to get the hearing for their moral claim to a larger share of the planetary wealth. Surely here the ethical basis for a new economic order does not come from thinking small.

As well, Schumacher does not address head-on the conflicting claims in any canon of distributive justice. My own canon, largely derivative from the economic ethicist John A. Ryan, allows a pluralism of claims to the distributive share of wealth. I argue that there are five items in an inclusive canon of distributive justice. An exclusive appeal to any one item in the canon would not yield justice. The five items or factors are:

1. Moral equality as persons, which grounds a principle of equality of opportunity;
2. Proportional need;
3. Efforts and sacrifice based on labor and exertion;
4. Contribution to final productivity in ways that do not depend on exertion. I am thinking of capital sacrifice;
5. Scarcity.*

I would also argue priority rules in cases of conflicts among these five factors. Needs have first priority to reward. Next, efforts and sacrifices are superior to productivity as claims to reward. Finally, productivity has priority of moral claim over scarcity.

Obviously, I do not have time here to justify my canon of distributive justice, its plural items and the priority rules for cases of conflict. But any widespread conversion to a more just and humane economic system or what Schumacher refers to as "a broad, popular movement of reconstruction," would seem to imply some growing societal consensus about the various factors which legitimately lay claim upon the wealth of nations and the planet and the weight

*For a developed treatment of the canon of distributive justice, see John A. Ryan, *Distributive Justice,* New York, 1927, pp. 213–221.

given to each factor in situations of conflict. I find Schumacher's implicit canon which seems to allow weight to need and exertion (labor) too narrow, as well as curious in an economist who, surely, would place some moral weight on contribution to final productivity in capital sacrifice (at least for small-sector capital which is not nationalized) and scarcity.

Finally, while I strongly applaud Schumacher's refusal to separate the issues of justice and liberty, once again he is not terribly helpful in suggesting how we sort out conflicting claims to social and economic rights, on the one hand, and civil and political rights on the other. It is just such conflicts between different kinds of rights which undergird contemporary policy debates about models of development. Some find the Chinese and Cuban cases unconscionable because they seem to neglect civil and political rights. Others oppose the Brazil case because it fails to recognize economic and social rights founded on basic minimal needs for food, housing, health, work, and so on. Those who argue for triage in food distribution give priority to potentiality and labor over need. Others champion the superior moral claim of need.

I suspect that the only way we can adjudicate these rights in conflict is an appeal to, first, foundational economic, social, civil, and political rights which are so closely connected with human dignity that they are, in all cases, inviolable. These rights need to be stated with specificity and, if necessary, in terms of their institutional embodiments. And then, there are second-level or derivative rights which are essential for humane development and freedom, although not so basic that they are always and everywhere inalienable. Jacques Maritain attempted some such sorting out of rights in distinguishing between rights of the human person as such, and rights of the civic person, in a book entitled *The Rights of Man and Natural Law*.*

I have the distinct impression that the contemporary policy debates about a new international economic order, development strategies, world food and population policy, and so on, will continue to be muddled until some consensus is reached on which rights to *both* bread and freedom are so fundamental that they may never be abrogated, and which are secondary, derivative rights which pertain

*Jacques Maritain, *The Rights of Man and Natural Law*. New York, 1943 (reprint edition New York: Gordian Press, 1971).

more to the developed well-being of a person than to his/her fundamental being as a person of moral worth and minimal needs. Obviously, in a case of conflict, the former would have priority of moral claim over the latter.

Sometimes in policy debates an appeal is made to protecting highly derivative civil or political rights (for example, the right to suffrage) as the moral basis for denying rather basic economic and social rights (for example, right to enough food to survive, a decent house, minimal health care, and so on).* On the other hand, these appeals gain credibility against systems of justice which, in principle, allow the sacrifice of fundamental liberties to achieve economic and social rights. The problem is that in a pluralistic world there is no consensus about what constitutes the core of fundamental inalienable rights and needs versus useful, second-level rights, and about priority rules for adjudicating rights in conflict. I am less sanguine than Schumacher that this consensus will grow out of some new metaphysics.

The social implication of neglecting this much-needed area of ethical analysis is that Schumacher's *Small Is Beautiful* will probably not be the basis for most policy discussions about a new international economic order, since its ethical analysis remains undeveloped in just those areas which underpin the value disputes in these policy discussions. More's the pity because Schumacher's is a wise and humane program of a man who knows much about both virtue and wisdom, two qualities in this world which are very much subject to the classic economic problem of scarcity.

---

*For a developed model for adjudicating rights in conflict, see Yale Task Force on Population Ethics, "Moral Claims, Human Rights, Policies," *Theological Studies*, vol. 35, no. 1, March 1974, pp. 83–114.

The Values of Appropriate Technology and Visions for a Saner World

Isao Fujimoto

The type of technology we have influences the kinds of values we live by and conversely the type of ethics we live by influences the kind of technology we will live with. What I'll speak to is this relationship between technology and values.

There have been a lot of changes in recent years that have profoundly altered the kind of values by which we choose to live and the way we look at the world. To start with some brief examples of how technology influences values: first, in the name of productivity and efficiency our technological approaches have changed society, to put it mildly. Particularly in America, we have seen a society characterized by an economy of scarcity changed to one governed by an economy of abundance. In the process we have not only changed materially, but we are witness to changes both ethical and spiritual.

When I was growing up during the Depression, there was this notion imprinted on all of us that we should save, not get things until we could pay for them, or to be more pious about it, delay gratification lest we tempt ourselves into satisfaction. This has dramatically changed. Instead of "Save Now," it's "Spend Now and Pay Later" (or forever). Along with this has come a peculiar notion about money and credit. We don't deal with money anymore; we deal more with credit. In the past, money was okay, whereas credit or spending what you did not have was not only bad, it was criminal. When I was a child I heard more stories about people who overdrew on their bank accounts than I did of murder and mayhem because not only were there heavy penalties for the former, but it seemed, from my childhood recollections anyway, that more people got put into jail for bad check charges than for mayhem. Today it is not this way at all. Banks encourage you to overdraw on your account and make it very easy for you to do so. As another example of how our values have been affected and reflect a move to a consumer-recruiting society, let's look closer at credit, which is a more sophisticated form of overdrawing on your bank account. The idea behind credit was to encourage the purchase of merchandise. Today it is the other way around. A review of the profit statements of major merchandisers, such as Sears, shows that income from installment interest is considerable, suggesting that merchandise is as much a means to sell credit as credit was meant to be a means to sell merchandise. Given such cases, we can't help but conclude that there have been some profound changes, including the changes of our attitudes and priorities on how we save, spend, delay, or sabotage our gratification. Such a major impact on our values didn't come about by breakthroughs in philosophy as it did in the impact of technology in our lives.

I think other countries are getting wise to us. Though the things we have here look pretty good to those in the countries who don't have them, their requests are tempered with certain qualifications. More countries are saying something like this: "We know you have a lot of good things, but we don't want the negative parts that go with them. We'd like your technology but we don't want the pollution, waste, and social cost that seem to go with it." I think this kind of expectation is unreal because technology, despite claims that it is value-free, neutral, benefits everyone and has no politics to it, is far from being all of this. The kind of technology that is transferred carries with it a lot of other baggage including assumptions about what is valued and what is not, who's going to benefit more and who less. Technological change, whether it be in agriculture, management, education, institutions, economic development, when transferred, transfers with it influences which benefit a few people at the expense of many, favoring one social class over another. Technologies can be class-oriented, be it the Bay Area Rapid Transit or the Green Revolution, when in both the short and long run, the people who benefit are skewed to the more advantaged social class. Also the kind of technologies which have been transferred in the name of progress and efficiency, carries with it characteristics of being exploitative of both people and resources. And lastly, related to its scale and capital- and energy-intensiveness, such technology leads to relationships and situations which are very non-participatory, undemocratic, and unecological. The characteristics of such a technology are far from being value-free, and in terms of its impact on how people relate to each other, the effect is considerable.

It might be good to stop and ask yourselves why are we where we are, and are there any alternatives to where we are going? One way to describe where we are is in terms of a technological paradigm that seems to dominate our lives. By a paradigm we refer to a way of looking at the world which influences what we do. Certain features characterize the technological paradigm which in turn shapes our behavior.

One is a notion of the technological imperative which says that any knowledge that can be applied should be applied. To hold back is to hold back progress. If left unquestioned, such a presumptuous view can lead to fatal consequences. An alternative to such an imperious view is one tempered by a sensitivity to the human condition. Those of you familiar with Pirsig's book *Zen and the Art of Motorcycle Maintenance* might recall an altogether different way of regarding technology. In Pirsig's view, "There is a Technology with a different face, technology that is not fragmented, but is connected to the heart and the spirit." Such an outlook encourages us to love all things as we love ourselves. All are a part of our lives. Everything around us, including technology and things, is dynamic. What's central to this distinction is the difference between quality and quantity. What we are more concerned about is the qualitative aspects rather than the quantitative ones. A quantitative look at Christmas for example is to regard this event as an extension of our social system measured in terms of the gifts, cards, or parties checked off or exchanged. But the whole point of Christmas, and many other kinds of holidays, is the focus on coming together to renew what's important as individuals, as families, as friends, as social groups. This is the qualitative dimension to Christmas.

Bringing it closer to home, I can share with you something I learned from my sons this week. This week my boys had friends staying over. Because I was working late and knew I wouldn't be up in time to prepare breakfast with them, I laid out all the breakfast ingredients on the table that evening. About the time I finally got up, one of my sons came to me and I asked him, "Well, did you have a big breakfast?" His answer was, "It wasn't big, because we didn't eat together," and when I heard that I thought I was staring in the face of Buddha, for I think he would have said something similar in pointing out the difference between the quality and quantity of relationships and events.

Another aspect about our being influenced by a technological paradigm is that of a technological fix, which says any problem created by technology can be solved by technology. For example, if a nearby river delta is considered too narrow, the response would be to dig it wider without considering that natural events would remind us not to consider such actions in the first place. Instead we get locked into technology as a way to solve our problems. An alternative to the technological fix would take into consideration the social dimensions and ecological consequences, lest solely technological approaches end in disaster.

A third feature of the technological paradigm is elitism. The structure of technology implies that the type of technology we have can only be handled by a certain stratum of people. This is another way of saying that technology influences our perception about the worth of people—that those who own, use, or build technology are more important than those who don't. I think this elitism is antithetical to a major idea behind appropriate technology. A basic premise of appropriate technology is that it is a technology in which all people can get involved—a technology more sensitive to people and the earth's resources.

What's being suggested here is an approach to technology in which instead of relying on experts, individuals can participate and become involved. It's a technology that is democratic and demystifies the idea of experts. There are consequences to the kind of technology we use. As suggested earlier, technology influences the kinds of values we live by and vice versa. Our being tied in to a notion of a technological fix is related to the way we measure how we are doing. We measure progress by a device called Gross National Product (GNP). To get a value for GNP, nearly everything is thrown in, including receipts from repairs on cars wrecked in highway accidents, the numbers of people treated (not necessarily cured) for incurable diseases, money made at bad movies, and so on. Everything accountable in money terms, whether it's a rationalization of our mistakes or bad cover-ups, is counted, and adding all this up tells us how much we are worth as a country. I recently heard Governor Jerry Brown remark on this matter of GNP and he said, "No wonder it's called gross."

Perhaps we ought to be thinking about a different way to measure the quality of our lives. Instead of GNP, I would suggest something called NWS. This means measuring what we are doing in terms of its contribution to Net World Sanity, rather than Gross National Product.

With this as prelude, I'd like to comment on the ethics of appropriate technology, comparing especially those values associated with either big or appropriate technology.

As an example, we can observe the alternative agriculture movement, which addresses the use of appropriate technology. There are many revealing signs as to how this movement is moving and who is in it. They are indications such as the number of seed packets sold, the shortage of canning lids, the number of urban gardens, people subscribing to *Organic Gardening and Farming,* and the number of people attending conferences like this. It is a very diverse movement, and yet despite this diversity all elements—be they people getting involved with family farmers, organic gardeners, farmworkers, food cooperatives, and so on—there are values that are beginning to emerge as commonalities.

I would like to review four basic values that appear to be shared by people who are involved in a move towards an appropriate technology and an alternative agriculture. The first is a premium placed on self-reliance or the idea of doing things on your own and having a better sense of control of both one's own life and what is going on. This contrasts with big technology which assumes that it is the experts who can do things for people. The second value shared by people involved in alternative agriculture and appropriate technology is the respect for decentralization. This refers to people being independent rather than dependent, self-determined, and valuing local control. Thirdly, there is more of an effort of people trying to do things as a group, things which they couldn't have done as individuals. Cooperation rather than competition is the mode. Fourth is the recognition of the importance of accountability. Accountability is thought of not just in terms of what's gained immediately but also for the long term; not in terms of maximum gains but more in terms of gains that are optimal. Rather than thinking in terms of just economic accountability, there is recognition that actions must be also socially and ecologically accountable.

Thus a comparison between the values of appropriate technology and big technology shows real contrast—as real as the difference between self-reliance and dependence, self-determination and centralization, cooperation vs. competition, and accountability as opposed to exploitation. These contrasts represent vastly different ways of looking at the world.

There is a relationship, and a direct one at that, between values and technology. We can't separate value from technology, and those of us who profess to work in one or the other end up fooling ourselves. If you are a scientist dealing with technology you also have to pay attention to the ethical considerations and the impact of that respective technology on the ethical aspect of our lives. Philosophy does not occur in a vacuum. Also, we have no choice but to recognize that we live in a world of material things which are manifestations of our values, and we have to deal with that reality too.

This leads to asking what would happen to the world if we in the United States put into practice the ethics associated with appropriate technology. To visualize this, perhaps we can put the realities of the world into the context of this gathering. There are at least a thousand of us here. Borrowing from an example cited from a Simple Living program, we can imagine that we, in this room, make up the entire world. As a global village of 1,000 people, 700 of us will be unable to read and only one of us will have a college education. Over half of us in this auditorium would be suffering from malnutrition, and over 800 would be going home to substandard housing. If we made up a global village of 1,000 residents, 60 would be Americans. These 60 would have half the world's entire income and the remaining 940 would have to exist on the other half. How would the wealthy 60 among you live in peace with those neighbors sitting near you? Surely you would be driven to arm yourselves against the other 940, spending, as much as we do in the United States now, more per person on military defense than the total per capita income of all the people in the auditorium.

If this is reality, consider what would occur if we started to act on the values associated with appropriate technology. For example, imagine the impact—not just on the United States but to people abroad—if we started acting on being self-sufficient. Despite cries that this would be impractical, as well as impossible, we need to remind ourselves that many countries are saying, and in no uncertain terms, that the best thing the United States could do to help them develop is get off their backs. This includes the effect of shipping technology—management, capital, equipment—that makes them beholden to a centralized system in the United States. A common notion is that we have six percent of the world's population but use over forty to fifty percent of its resources. This means that our standard of life is obtained at the cost of other people's welfare. By thinking not just in terms of efficiency, but also accountability and cooperation, with ourselves and with others, we can dramatically affect the course of both human and ecological events.

I want to close with a statement of how values and technology come together. Those of you who have been buying bicycles from Japan may want to take a closer look at the repair manual that came with them because some carry a very appropriate statement. The statement goes something like this: "Repair of Japanese bicycle require great peace of mind." You might wonder what that has to do with technology, and I will say that it has everything to do with it because what peace of mind entails is the integration of ourselves with our surroundings. Peace of mind deals with how we relate harmoniously with everything we do. The gist of appropriate technology is a concern not so much with the quantity of things as with the quality of life; about caring and relating to each other and the earth's resources much more wholly. I think all of this is part of a more meaningful way to look at the world, guided by values such as self-sufficiency, decentralization, cooperation, and accountability. These values are integral to this vision of a new or saner, more human world. Appropriate technology is a part of this saner, more human vision.

The Ethics of Thinking Small

E. F. Schumacher

I will tell you how it all started. I noticed that the scientists and technologists of the rich societies attended to the problems of the rich societies, while the scientists and technologists of the poor societies also attended to the problems of the rich. This applies to the United States and India, but it also applies internally between São Paulo and the northeast of Brazil. The technologists and scientists everywhere tend to devote themselves only to the problems of the rich. I also noticed that the problems of how the poor could rehabilitate themselves were becoming ever bigger. There we were, decade after decade, trying to help the poor countries—with what success? There are now more poor people in the world than ever before. The gap is not narrowing. Why? Even in the rich countries the gap is not narrowing. Maybe the proportions change a little bit, but we are not making a good job of it. Even in this country I hear it stated, from time to time, by your presidents, that twenty or thirty or forty million people are living in misery. Why? Why? The theory is that we can eliminate misery only if we have more "economic growth." But if America, with an average income per head twice as high as that of Western Europe, has tens of millions of people living in misery, what hope is there? Besides, the Club of Rome is telling us—not without good reason—that there are limits to growth. To drive for "growth" in the hope that wealth created among the rich will *percolate* to the poor is evidently unrealistic. Wealth never percolates; it stays with those who have acquired it in the first place. The only way to combat poverty is by enabling people to help themselves. How can people help themselves? By becoming more "productive." And what are the means towards higher productivity? Technology.

I noticed that there is always available a very primitive technology which is generally not good enough to produce a tolerably good living standard, and there is always available a very "high" technology which is totally outside the reach of the poor. In between, there is virtually nothing. The "middle" has disappeared. So if we want to help the poor to help themselves, we must fill that middle.

I am not an academic, I'm a businessman, if you like—a man of action. I run a number of organizations. So, I don't write books that mention everything; everything I write arises out of the work that I am actually doing. And I apologize to the professors that certain things are not mentioned in my book that they are very interested in, but if I mentioned them all then we would have no need for professors.

There is this strange phenomenon of losing the middle that you can observe in every book shop. You can get the latest and the best, and you can get the classics, but the middle has dropped out, it's out of print. When I was a farm laborer in England some thirty-odd years ago, the total equipment for a farm would cost a few thousand dollars. And so, all sorts of humble people could be farmers. Now, that equipment has disappeared. It is no longer there. Nobody makes it and if you want to farm, you've got to be a rich man, because the equipment for the same farm now costs a couple of hundred thousand dollars. The technology has developed to exclude the little people more and more. At present, in the Western world, we have fifteen million unemployed, and they can only sit back and wait. You may remember a play, *Waiting for Godot,* and the two tramps under a withering tree and Godot never comes. So in England they are sitting and waiting for Whitehall, and here they are sitting and waiting for Washington.

Well, I don't think that's the way to deal with the matter. One can't forget about fifteen million people. Is there a technology in existence or do we find it in our hearts to develop a technology so these people can, on a small scale, rehabilitate themselves? The answer is yes, of course it can be done, if we find it in our hearts to do it. And if people say to me, "They," the big corporations or goodness knows who, "They won't let you," well, I can only reply with Chairman Mao that these are paper tigers: as far as intermediate technology is concerned, they are paper tigers.

In any case I never deal with corporations. I only deal with people. And I actually have found very able people even inside the big corporations. Even inside government and even inside academia. In other words, these are excuses for doing nothing. Sitting back waiting for daddy, or the state, will not solve problems. The subject of self-reliance has already been hammered once or twice, to my delight. We actually don't have to wait for anyone. We can get down to it, we can have a technology that we can afford, that even small people can afford. We don't have to leave it to the multinational corporations. What will become of them? I don't know. I have found that they can well look after themselves. But I've also found that the poor can't look after themselves very well, because the system denies their right of existence. So this movement of intermediate technology is a movement to try and to utilize our splendid knowledge, our great technological ability, in a direction that doesn't make the gaps bigger, but smaller.

And how is it to be done? As a practical man I know that you can't suddenly make a turnabout. That's not the way change happens. Change must come from the periphery, gradually. So I would say, carry on with all your little games, moon landings, and the lot, but occasionally think of the little people. Perhaps five percent of research and development expenditures should consciously be utilized to develop a technology whereby small people, small companies can help themselves, and whereby the richness of life can be restored in the more outlying areas of the United States, in the outlying provinces of Canada and, in fact, all over the world. These areas, everywhere, are crying for help. "We are becoming a colony of the big city; if there's a possibility of a small-scale technology, please come and help us." This problem is no longer confined to the Third World; it is world-wide.

Everywhere one of our most urgent tasks is to find the *appropriate scale* of our technologies, our organizations, even our political units. If things become too big, they become inhuman, even antihuman, and sooner or later inefficient and counterproductive. They promote various processes of polarization, into the very rich and very poor; into areas of vast congestion and areas of vast emptiness. They lead to the destruction of all harmonies. Life is a matter of relationships. This sets a limit to the size of groups—if the groups are to remain human. In a group of a dozen people, you have 66 bilateral relationships, which still makes it possible for each person to know how everybody gets on with everybody else. Some of the greatest teachers of mankind have found that a group of about twelve is about right. Research people have found the same. If groups are bigger they tend to split up into "primary work groups" of about twelve, and this often leads to very unsatisfactory results. In a group of a hundred people there are 4950 bilateral relationships, which is a very difficult matter to carry in one's head.

After a century or two of hammering the slogan, "The bigger the better," and now finding our most vital problems becoming virtually insoluble, we shall be well advised, just in order to restore some sort of balance, occasionally to entertain the thought that small may be beautiful.

Discussion

AUDIENCE PARTICIPANT: This question is addressed to Dr. Coleman. He stated that there was the argument that there is no moral basis for the claim of future generations upon this generation. I suggest that there is. I ask him why he cannot recognize that we are in a moral crisis, based upon our crisis of survival, and that the old ethic will no longer work. Why do we not have a new ethic, a new ethic of the survival of the species as the highest good?

JOHN COLEMAN: I want to make it very clear that I did not claim that there is no moral basis by which future generations can lay claims to us. Far from it. I know intuitively that that's correct.

But what I was trying to say was that intuitive notions of ethics are not always helpful. On the basis of that intuition, let me point to two different policy conclusions. One is, we do not have to attend to human rights presently in Brazil, because the peasants there have to make sacrifices, and costly renunciations in the name of two generations from now. I find that somehow I know something is wrong. On the other hand, there are claims that because of future generation moral claims on us, we must be responsible for terms of world population policy. That seems intuitively correct. What I was calling for was some attention to the grounding of those moral claims. That is, the basis on which that claim is both made and limited must be grounded. Surely the future generations do not have unlimited claims on us such that there can be a total sacrifice for economical miracles in Brazil and elsewhere of this and future generations in the name of simply some future good. It's clear to me that there is a moral claim, but all I'm asking for is a lack of ethical muddle-headedness. That will come only if we attend to the basis on which we ground that claim. Otherwise, we get intuitive notions which don't help us to adjudicate rights. In fact, in most policy discussions everybody has a moral claim. The question is, which takes priority in conflict, what are your principles for justifying that?

AUDIENCE PARTICIPANT: How do we move in a direction of intermediate technology if, in getting there, we stop growth and go through a world of social collapse and bankruptcy?

E. F. SCHUMACHER: Well, I shouldn't worry too much about it. It's only on the money side, and that's not the real side. It's quite easy if the debtors don't pay. I'm not so terribly worried. If people have too much debt, they ought to default. And the creditors will be extremely angry and call them names, but life goes on. Now, of course, if these things are taken that seriously, then there may be just a general confusion and a depression. That can also happen. But the way I see Britain raising loans, it seems to be much easier than default. It's not my interest, quite frankly. I would like to stick to real things. To the hungry people, to the work opportunities. On the whole, that has very little to do with these games played in high finance.

AUDIENCE PARTICIPANT: On page 54 of *Small Is Beautiful* you describe the threefold view of the function of work. How would you use this approach to begin an appropriate definition of the quality of life?

E. F. SCHUMACHER: Well, working in Africa we developed some industrial activity. We trained applicants and it all worked very nicely. We had the appropriate technology, as we thought. The only troublesome thing was that the applicants so quickly disappeared again. And normally, one couldn't find them. But occasionally we would get hold of one and say, "Where are you, why didn't you come back?" And the applicant would laugh. He'd say, "I'm a man, I'm a person. It was nice for four weeks to do the same thing all the time, but it's not a real person's life." So, let's not forget this.

Most of our masters have never been industrial workers, so they carry the suffering of others with great fortitude. But this person was too big a man, a real person to suffer the type of industrial technology we tried to introduce into Africa. Now again, in our countries, we found that many, many jobs are simply below the level of the unfortunate fellows who are supposed to do it. And so they become, of course, very unhappy. Very unhappy, and often they just drop out altogether. As Thomas Aquinas has mentioned, the joy of work equals the joy of life. You can't get it in the later hours by watching television. He also said indolence is associated with sadness of the soul. If there is nothing real to do but a few motions of highly organized production, I think this era is coming to an end. I wish it would come to an end. I wish our ingenuity would go into producing a technology which is worthy of persons. In most cases it's unworthy now. It's just another stage of slavery.

AUDIENCE PARTICIPANT: It appears that tools develop and become complex machines and then all the machines in the middle die off. It appears that work has a tendency to become degraded, all by itself, as if there weren't a human, social process that was making it happen. Why did it happen? Did it happen because of certain social relations?

DESMOND JOLLY: I think perhaps the panelists were addressing themselves to probably some more fundamental questions. As I said, it seems to me that the kind of alienation and oppressive bureaucracy in large-scale systems we are talking about could obtain, and probably does obtain, in the social system itself. So I think we are probably dealing with a fundamental issue.

E. F. SCHUMACHER: We are all searching for the right social system. One way of changing the social system is to create new possibilities of living for those who don't get the benefit. I was in East Germany, a Communist country, not so long ago, and they were very, very critical of the West. They said, "The West, well of course we are much more efficient than they are. They are like an express train hurtling at ever-increasing speeds towards an abyss, but we shall overtake them."

AUDIENCE PARTICIPANT: Dr. Schumacher, in a few sentences, could you clarify, especially as it pertains to say a General Motors factory in Fremont, how control at that level would relate to say the general corporate structure of General Motors, which is headquartered in Detroit?

E. F. SCHUMACHER: It has been a long time since I worked for General Motors. You'd better ask them. All I can say is that it doesn't seem to me to be a very clever system to say, now you go ahead and make an income and then I, the state, after you pocket the income, will try to get 50 percent out of your pocket. Because when I have it in my pocket, I'd like it to stay there. Now, the state requires money to be able to work in a well ordered fashion. It's a tremendous blessing, as anyone knows who has tried to do anything productive in a not well ordered state. So, as far as the big corporations are concerned, surely to say that this is private enterprise to be carried on for the benefit of the shareholder's full stock is just a fiction. It can't be. There must be social influence. Can't we bring in the social influence in a more elegant way by incorporating it into the ownership structure? I wrote that if you have a certain capital in so many shares, then double them and half of it goes into some public hand. I don't mean the federal government, but something very local. This gives the social factor, the political factor some rights. But these rights should be normally held in abeyance because bureaucracy and business don't mix very well at all. You may get a bureaucratic business or corrupt government and so on. But the public eye should be on exactly what they are doing. The representatives of the public should be able to say, "No, you can't do that." These decisions should be taken in the full light of public responsibility or social responsibility.

I suggest one scheme about how this can be implemented without any bloody revolution. It would alter the tone of society. It would not be necessary anymore for little journalists and gadflies to try and ferret out what the bulk of the immensely powerful corporations are doing, because the public, as a 50 percent owner, without normally interfering in the management, would be there and would see what is happening. I spent twenty years in a nationalized industry, and I think to bring this in doesn't interfere with the efficiency of management. It would be very, very healthy. Also, it would act to close the credibility gap which now exists between the large corporations and the people at large.

INTERMEDIATE TECHNOLOGY, THIRD WORLD DEVELOPMENT, AND NEW MODELS OF CENTRALIZATION: IMPLICATIONS FOR THIRD WORLD PEOPLE IN THE UNITED STATES AND ABROAD

Lyman P. van Slyke

Alex F. McCalla

Garrett Hardin

P. K. Mehta

INTRODUCTION: *Yvonne L. Hunter*

MODERATOR: *Donald R. Nielsen*

Introduction

Yvonne L. Hunter

While some individuals may question whether intermediate technology is suitable for developed nations, most agree that intermediate technology has considerable promise if applied correctly in Third World nations. Indeed, many proponents claim that the advent of intermediate technology is the only salvation for these nations.

Put very simply, Dr. Schumacher maintains that in developing nations, energy-intensive, capital-intensive, and labor-extensive technologies are not appropriate, given limits on capital, water, and land; the population levels; and rural orientation. The case for an intermediate technology approach in developing nations is stated very eloquently in *Small Is Beautiful* when Schumacher speaks of the intellectual challenge of development:

> The aid-givers—rich, educated, town-based—know how to do things in their own way; but do they know how to assist self-help among two million villages, among two thousand million villagers—poor, uneducated, country based? They know how to do a few things in big towns; but do they know how to do thousands of small things in rural areas? They know how to do things with lots of capital; but do they know how to do them with lots of labour—initial untrained labour at that?*

This, then, sets out the challenge of intermediate technology as Schumacher sees it.

---

*E. F. Schumacher, *Small Is Beautiful: Economics As If People Mattered,* 1975, p. 196.

Many of Schumacher's critics contend that the intermediate technology solution is not the answer, since it does not speak to the need for economic growth. "But without economic growth [in developed nations]," writes Burton G. Malkiel in the *New York Times Book Review,* "how can we ever provide minimally decent living standards for the billions living in the less developed world?" * In a similar vein, other critics assert that the primary goal of technology should be to maximize output—to produce the most with the least—so as to meet the urgent human needs as fast as possible.

Proponents of appropriate technology reply by pointing out that growth is not necessarily the answer, since it cannot bring harmony between rich and poor nations. In addition, they point out, as indicated in an article by Wakefield and Stafford writing in *The Futurist,* that modern mass-production technologies have not solved the problems of unemployment and that they have often resulted in goods and services being too expensive for average people to buy.†

Another charge often raised by critics, and described by Wakefield and Stafford, is that appropriate technology constitutes technological imperialism. "In the view of such critics, appropriate technology means second rate technology and stems from the desires of developed nations to hoard their most advanced technical devices in order to discourage competitive technological development elsewhere."

Schumacher and his followers would certainly disagree with this criticism, but it still remains a thorny issue which is exacerbated by the involvement of the United States Agency for International Development (AID) in the expansion of intermediate technology projects in developing nations. AID has created a non-profit corporation called AT International which draws its support from funds appropriated by the United States Congress under the Foreign Assistance Act. The charge of AT International is to "expand and coordinate private effort to stimulate the development and dissemination of appropriate technologies in developing countries."

Thus, it seems that the spread of appropriate technology in Third World countries is carried on at two levels: the individual or small

*Burton G. Malkiel, "What to Do about the End of the World," *New York Times Book Review,* January 26, 1975, p. 19.

†Rowan A. Wakefield and Patricia Stafford, "Appropriate Technology: What It Is and Where It Is Going," *The Futurist,* April 1977, pp. 72–76.

group, such as Schumacher's Intermediate Technology Development Groups, and the larger national organizations like AID. In many cases, misunderstandings about the motives of *all* groups occur because of the sometimes questionable activities and motives of organizations like AID. This was evident at the conference by the numerous questions concerning Dr. Schumacher's involvement with AID. While it is not appropriate here to discuss the "morality" of AID's alleged activities, suffice it to say that AID's involvement in intermediate technology projects should not be used by critics to discredit the activities of other appropriate technology groups.

The four panelists in the Third World session cover a wide range of ideas.

Lyman P. van Slyke cuts through the often romanticized references to China as a model for Third World development and presents a detailed picture of small-scale industry and intermediate technology in China. He also discusses the long-range prospects for this strategy to continue in China.

Alex McCalla presents a critique of the Third World section of *Small Is Beautiful.* In his talk, he draws specific critical attention to some of Schumacher's points, but in no way detracts from Schumacher's main arguments.

Garrett Hardin maintains that we must be careful not to harm developing nations, even though our aim may be to help them. He discusses the nature of work and how it relates to Schumacher's ideas, which he says produce a positive externality.

In his discussion of India, P. K. Mehta concludes that the adoption of intermediate technology and family-planning programs cannot solve the problems of poverty unless the social, political, and economic institutions are restructured to become more responsive to human welfare.

When examined in conjunction with the other thoughts on agriculture, these panelists provide a powerful and positive statement about intermediate technology and Third World development.

Rural Small-Scale Industry in China

Lyman P. van Slyke

The remarks that I want to make about present-day rural small-scale industry in China and its future prospects do not come from my experience as a historian, since historians are usually better at telling you about the past than the present, and they are often wary of predicting the future. Instead, my knowledge of Chinese economic development and small-scale rural industry comes largely from my participation in a delegation which visited China for a month in the summer of 1975, and looked specifically at rural small-scale industry. (The report of this delegation will be published in July 1977 by the University of California Press,* so if you are interested in what China is doing in this area, you will have a chance to check my remarks against the reactions of our group as a whole.)

I want to begin with the context, the general background, within which rural small-scale industry and intermediate technologies are employed in China. One of the most important elements of that background is that China sees herself very much as an industrializing and modernizing society. Chou En-lai, in his Political Report to the Fourth National Peoples Congress in January 1975, sounded what is still today the marching order for China. He said:

> The first stage is to build an independent and relatively comprehensive economic and industrial system before 1980. The

*Dwight Perkins, ed., *Rural Small-Scale Industry in the People's Republic of China,* Berkeley: University of California Press, 1977.

> second stage is to accomplish the comprehensive modernization of agriculture, industry, national defense, science, and technology, before the end of this century so that our national economy will be advancing in the front ranks of the world.*

Now, while I think nearly all Chinese agree on this goal in general terms, they do not all agree by any means on a single strategy for achieving it. Much of the political struggle which we have witnessed in China—during and since the Cultural Revolution, and in very recent months with the elevation of Hua Kuo-feng as chairman of the party, and the purge of the so-called "Gang of Four"—involves among other things, conflict over questions of economic development and economic policy. For example, what should be the relative balance between agriculture and industry? How much stress should be placed on capital-intensive industry as against intermediate technologies and handicrafts? What about rural versus urban development? How much emphasis should be placed on further industrialization of cities and how much should be placed in the countryside? Should industrial development and economic development more generally be concentrated in areas that have comparative advantage in terms of trained manpower and other resources, or should they be distributed with a view to bringing backward areas forward, even though from a strictly economic standpoint this might be less efficient? To what degree should China employ advanced and capital-intensive methods as opposed to labor-intensive intermediate and low technologies?

One must realize that these are also political questions, and that China's economic policies are strongly politicized. Economics not only impinge on politics, but politics impinge as well on economic problems. I do not mean to suggest that this is a monolithic society. Quite the reverse. It is a very diverse and very complex society with considerable initiative at various levels within it. But at all levels, from the local level of the work team and the brigade, the commune, up through county level, the provinces, and right to the center at Peking, politics are very important. And whether on the economic side or the political side, issues are not closed, because there is not a single strategy upon which all agree. Therefore, I

---

*Chou En-lai, Political Report to the Fourth National People's Congress, January 1975, quoted in *Peking Review* no. 875, January 24, 1975, p. 23, and often reiterated in recent months.

believe that China does not present us or the rest of the world with a clear-cut model of development. There is, within China, I believe, a range of competing models; I think we would do well to keep this dynamic and open-ended kind of perspective on China as we observe what she is doing now and will be doing in the future.

Now, having said this as background, what can be said about small-scale industry and intermediate technology in China? The Chinese themselves frequently say that China must walk on two legs, that she must somehow combine the dichotomies that I suggested above. It is not *either* advanced *or* low technology, but *both* in every area: urban development *and* agricultural development, industrial development *and* agricultural development, and so on. In one sense, of course, this simply poses these issues in a different form without fully resolving them. It might be well, then, to turn to some of the more specific elements and aspects of a small-scale industry in China.

First of all, what do the Chinese mean when they say "small-scale"? This is a fairly precise term. It means industry which is managed at the county level or below. It is either by county or commune, which is the next organization level below it, or by the brigade, which is the next level below the commune. Some "small" industries, particularly those managed at county or commune levels, are fairly sizable, employing in the plants we saw up to 550 and 600 workers. Others, mainly at the brigade level, are much smaller and more primitive; the smallest plant we saw was a simple machine shop with six workers. Thus, small-scale is to some extent an organizational term.

In recent years the principal emphasis for small-scale industry in China has been what the Chinese call "the five smalls," the five kinds of small-scale industries that are officially encouraged. Naturally some variation exists from place to place and not all of these industries can be developed in every area, but counties that can support "the five smalls" are pushed to do so in most cases. Furthermore, these five industries are meant to support agriculture directly. This is not industrial production for the national market or for the export market, but for local use. One of "the five smalls" is energy, especially hydroelectric energy, which ties in very closely with the whole effort for water conservation, since many small dams, reservoirs, and irrigation canals are also used for generation of hydroelectric power. There is also some small-scale coal extraction which falls under the general heading of energy development. A second component of "the five smalls" is chemical fertilizer, or what more generally might be called agro-chemicals, including not only fertilizer itself, but also pesticides and explosives. Explosives are chemically related to fertilizers, and are used for blasting and for the development of water control, irrigation, and so on. Cement is yet another of "the five smalls," not only for building construction, but also for what the Chinese call "farm land capital construction", for lining causeways, facing dams, and so on. The fourth element is smelting of iron and steel, where possible. Many people must still remember the lurid pictures of the backyard furnaces during the Great Leap Forward, where nearly every bureaucrat or farmer was expected to stoke his own small blast furnace. This has now been much rationalized, and many areas do not have the natural resources to support a small iron smelter. But where local deposits exist, they are used.

The last of "the five smalls" is machinery, agricultural machinery primarily, which includes a number of kinds of activities. To some extent, this involves the manufacture of machinery itself, either the entire piece or substantial components. For example, one of the most commonly seen pieces of farm equipment is the hand tractor, which looks like a large rototiller. It is driven in the field by a person walking behind it, although usually there is an arrangement which allows it to be hitched to a trailer for over-the-road transport as well. This particular design is common in many parts of Asia, not simply in China. Now, at the small-scale level, it is impossible for these relatively small machine shops and machinery plants to manufacture all parts of the tractor. In general, they do not manufacture the engine, the headlights, or the tires, and these are imported from other parts of China. But the transmission and the sheet-metal work and many of the other components may well be manufactured at the small plants. Water pumps of a variety of types, both gasoline and electric, are often made in such plants, as are a variety of other farm implements, right down to simple hand tools. In addition, in many of these shops, a portion of plant capacity is used to build machine tools. That is, some lathes and drill presses were being used not to make the farm machinery but to make additional lathes and drill presses. These plants were thus increasing their own future capabilities at the local level. Equally important is a machinery-repair capability. It is crucial, in a country where there isn't a Ford agency just down the road, that the local unit be able to maintain and repair its own equipment. Indeed, in the busy agricultural season many small farm machinery plants close down temporarily, and the work force forms mobile repair units that go to the fields with spare parts and tools in order to repair equipment on the spot.

Finally, a very important element is the training function played in all parts of the small-scale industry spectrum, but particularly in the machinery plants. Countless times we saw two people on a machine. One was a journeyman, the regular worker, and the second was an apprentice, a younger person, often a young woman, who was learning to operate the machine. These are, then, "the five smalls":

1. energy, especially hydropower;
2. agricultural chemicals, especially fertilizer;
3. cement;
4. metal, especially iron and steel; and
5. agricultural machinery, including manufacture and repair.

What are the special advantages or goals at which small-scale industry is aiming and what can it achieve? It certainly looks to a more evenly balanced development and to a degree of local self-sufficiency. In China, self-sufficiency or self-reliance does not usually refer to 100 percent self-reliance, but to a situation in which the state helps those who help themselves, while they do as much as they can locally, in terms of mobilization of labor, mobilization of capital, and mobilization of other resources. Then, if there is input needed from higher levels, that can be supplied. This approach relieves large-scale industry of many tasks which it would otherwise have to perform or leave undone. For example, if it were not for small-scale nitrogenous fertilizer plans, using in many cases fairly low-grade materials, the large plants would have to supply that need—if they could. Similarly, because some capital is accumulated locally, higher levels are relieved of a significant investment drain in industry and the tasks of collecting and reallocating resources are eased. And, very importantly, it relieves the transportation system, which is still very much underdeveloped in China, of many tasks it could hardly perform. If there is a small seam of coal in a locality, it is more economical to mine that coal even by very inefficient methods than it is to import coal from large mines by rail from an area only fifty or sixty miles away. Since the transport system is one of the critical bottlenecks in the entire Chinese economic pattern, the same economies apply to other high-bulk, low-unit-value products like cement and chemical fertilizer.

Moreover, since local industries meet local needs, they remain flexible and responsible. We observed this flexibility on many occasions. The engines for hand tractors are often of stock manufacture, but locally fabricated accessories, tire placement (to fit between crop rows), and so on, can be adapted to local conditions. Flexibility is also enhanced by the use of batch production rather than continuous flow methods, even though the latter are seemingly more efficient. A machine shop can manufacture water pumps, for example, until enough have been produced, then shift quite easily to some other task. Such a shift is not easy for a large plant using assembly-line techniques.

So this kind of industry is relatively flexible. These kinds of industry can utilize resources which would be too scattered or of too low a quality to warrant their being concentrated at large industrial sites. This is true not only of mineral and material resources, but also of capital and labor, some of which is seasonally available during the slack season in the agricultural calendar. And, of course, I have already noted, small-scale industry is an important educational and vocational training institution. Many young people are getting a firsthand education in the use of machinery, which in terms of large plants is quite primitive, but which is still fairly sophisticated in its own right. This is very similar to the kind of experience that many Americans went through in the nineteenth and early twentieth centuries.

How much production do "the five smalls" account for in China? In terms of China's gross industrial output, about six percent, according to constructed estimates from the outside, since the Chinese do not make this kind of economic data available. Within the specific areas of the five small-scale industries, they account for a much larger percentage, at least one-third of the total in these areas overall. This ranges from a high of about sixty percent in chemical fertilizer and cement, down to about five percent of the electric power, and ten percent of the iron and steel. Output is thus unevenly distributed, but in certain areas, particularly in chemical fertilizer and in cement, it accounts for a substantial part of total national production.

When all of this has been said, we are still left with an unanswered question. In fact, I believe that China is left with an unanswered question: will small-scale industry and the use of intermediate technologies become a permanent feature of Chinese development, or is this a medium-term expedient forced on China by present circumstances? At her current stage of development, the use of small-scale industry is rational and its results are impressive—though perhaps less impressive than the most optimistic observers would have us believe. But what of the future, when present shortages and imbalances have been substantially overcome, when transportation bottlenecks are no longer so severe? Will China then see small-scale industry and intermediate technologies as ends in themselves? Those responsible for small-scale industry in China, and rightly proud of the accomplishments of their particular plant, told us over and over again that their goals were to move "from small to large, from primitive to modern, and from here-and-there to everywhere."

My own guess is that small-scale industries and intermediate technologies will continue to exist for some time, because China is such a large and diverse country that the conditions rationalizing this effort will not soon be transformed. But if China continues to define her goal as "advancing in the front ranks of the world," and if she defines what that means with the same terms and standards used by the so-called "advanced nations," then intermediate technologies may face an uncertain future. Neither we nor China can yet know whether "small is beautiful" or simply useful for the time being.

The Third World: Small Is Beautiful

Alex F. McCalla

In the brief remarks that I wish to make at the outset I shall confine myself to Part 3 of Dr. Schumacher's book, titled "The Third World." I shall emphasize mainly those elements that relate to development in the low-income developing countries. I begin by briefly identifying what I consider to be Schumacher's major points. I then turn to commenting on each of them and conclude with some general comments. I should state at the outset that in this particular section of his book I find much more with which to agree than to disagree.

The process of development involves the totality of activities that relate to the social, economic, and physical well-being of masses of poor people. As such, development is not simply a program of aggregate national economic growth. In fact, it is not even a totally economic issue—it is also a social and human issue. Thus, approaches to development must focus on people and their needs. Development is therefore concerned with education, organization, and discipline more than it is with things. Given that two-thirds of the world's population lives in developing countries and that the majority of these people live in a rural agricultural environment, the approach to development must be decentralized and adapted to diverse specific needs. Further, given population growth rates in developing countries, the issue of increased gainful employment in rural areas is of prime importance. The short resource in developing countries is capital; the abundant resource is people. Therefore, development efforts should attempt to make productive and meaningful use of the abundant resource, that is, maximizing work places within capital limitations. This implies a small-scale, labor-intensive, rural-oriented development strategy. Such an approach is difficult to sell to people from developed countries whose own development has been premised on the opposite approach. It is also difficult to sell to educated city folk from developing countries who aspire to wealth following the western pattern.

I trust that this is a reasonably accurate reflection of some of Dr. Schumacher's views. Now I will turn to my own comments.

The concept of development as a comprehensive enterprise involved with the totality of people's needs is an essential concept. Unfortunately, as Dr. Schumacher points out, the demands of decision makers for quantitative analysis have skewed the attention of development workers toward those things that are objectively measurable (for example, physical and monetized economic variables). I don't despair quite as much as Dr. Schumacher about the absence of a commitment to total human development. It is my judgment that the concept of total development is increasingly accepted. The real challenge is to put it into operation. Here, the desire to telescope time, and therefore "create" rather than "evolve," has put immense pressure on policy makers to seek quick, simplistic solutions which, invariably, involve the importation of someone else's experience. The real issue is to develop indigenous problem-solving capacity at the local level. Here, the strong technological scale-biased approach to education and research in developed countries has great limitations.

The plea for attention to rural employment creation is well taken. If one puts pencil to paper and computes some simple numbers with respect to available labor force in the future, one must conclude that if people in developing countries are going to be employed, increasing numbers must be employed in agriculture and rural areas. This is true even if effective population policies are put in place immediately. If a nation has more than fifty percent of its current population living in rural villages and the population is growing at more than two percent per year, then, even with the highest possible rates of increase in urban employment, employment in rural areas must increase by at least fifty percent in the next thirty years if there is to be anything approaching full employment. This is a message which is not yet fully appreciated by many economists and others who have an urban industrialization bias on economic development. If a nation is to accomplish increased rural employment, of necessity it must follow a decentralized approach using limited capital to create as many jobs as possible.

In regard to this crucial point let me make several interconnected comments.

First, the case for a rural-oriented, decentralized approach to growth is forcefully made by John Mellor in his new book *The New Economics of Growth.** This treatise, amply buttressed by empirical evidence from India, makes a strong case for emphasizing agriculture and rural-related industries as its focus of development strategy, as opposed to traditional industrial leading-sector approaches. Some economists at least have seen the light. Secondly, the World Bank has at least changed its rhetoric with respect to its emphasis on the rural poor. Whether in fact the character of bank loans, which traditionally have emphasized capital works projects, will change, remains to be seen. No doubt one difficulty the bank will have is that it must deal through national governments which are urban-based and oriented towards urban industrialization.

A related point is that within national governments, a real conflict exists as to whether agricultural policy should foster maximum food output to satisfy the hungry in the cities, or whether it should focus on increasing the output of small, poor farmers who are much less politically powerful. The policy obviously must be concerned with both.

There is one point where we must be careful not to overdraw Schumacher's conclusions. Schumacher makes a strong point against capital-intensive machine technology, and rightfully so. However, there is a danger that we will apply the stricture against all technology. Not all technology is machine technology nor does it have to be biased in terms of scale. A new variety of cassava which increases yield is not necessarily scale- or capital-biased. A problem, however, may arise in that access to the complementary resources necessary to implement the technology may, in fact, be size-biased. This latter problem is not a problem of the technology per se, but of the social structure surrounding its use. It is important that this crucial distinction be kept in mind.

It should also be noted that the use of technology should be

---

*John W. Mellor, *The New Economics of Growth: A Strategy for India and the Developing World*, Ithaca, N.Y.: Cornell University Press, 1976.

appropriate to the particular environment. All mechanization may not be bad. For example, in specific instances, mechanical water lifting or land preparation may break bottlenecks and lead to *both* increased production and employment. A recent study by Raj Krishna in India delves into this point.*

Finally, we must keep clearly in mind that a rural-oriented employment strategy need not be limited to agricultural production. There are many other small-scale activities possible in the rural community.

These comments are made to draw specific attention to some of Schumacher's points and in no way to detract from his major argument.

I'd like to make one final point of agreement with Dr. Schumacher. He emphasizes time and time again that development strategy is essentially an attitudinal problem. If attitudes are conditioned by Western experience geared to maximizing physical output without regard to distributional issues, then the kind of approach he advocates will have difficult times. However, my limited experience suggests that in the rural areas of developing countries there is a growing cadre of dedicated people who are convinced of the necessity of a change in approach. This, coupled with an abundant supply of entrepreneurial talent in developing countries, suggests that some changes may be coming.

---

*Raj Krishna, "The Measurement of Direct and Indirect Employment Effects of Technological Change in Agriculture," in Lloyd G. Reynolds, ed., *Agriculture in Development Theory*, New Haven, Conn.: Yale University Press, 1973.

Human Resources and Third World Development

Garrett Hardin

By this time you should have a pretty good idea of what is in Schumacher's book, even if you haven't read it, because we've had some good summaries of it. It's a splendid work that points the way to a future we should work for. Let me take up one issue that falls into the category that the economists speak of as "externalities," or consequences not directly aimed at by a particular measure. This particular externality, I'm happy to say, is a positive externality. You don't often find these. Most externalities are a minus.

A positive externality of Schumacher's program is connected with the nature of human work, the allocation of human energy. When I speak of human energy, I don't mean anything mystical or non-physical, I just mean the energy from the environment that happens to get channeled through the human body. It goes in as an input and then comes out as an output, and the output is usually what we call work. I think that the nature, amount, and distribution of work is at the heart of the political problems of all modern states, rich or poor. In our effort to help poor countries, of course, we give them what we regard as the best information. Unfortunately, it often turns out not to be the best information because we ourselves are deceived. In this regard I think we are in the grip of some crippling superstitions. A major superstition of the European industrial civilization might be called effort quietism, or the belief that the normal human being prefers inaction to action. This belief is well expressed in what Keynes gave as the traditional epitaph written for herself by the old charwoman: "Don't mourn for me friends, don't weep for me never, for I am going to do nothin' forever and ever." Well, for one who has labored too hard and too long at too boring a task, this epitaph is quite understandable. But there's a wealth of information from animal behavior, from human physiology, and from anthropology that tells us that the truth is quite otherwise. The normal state of all human beings is activity. Not all of the time and not at top speed, but at a fair clip most of the time. Human beings are, of course, not unique in this. To see that all you have to do is go to the zoo and look at the monkeys for a while.

Once the inextinguishable urge to be active is recognized, we realize that we must take a second look at the policy of inventing, producing, and marketing labor-saving devices. In fact, one might ask, is not the purpose of promoting labor-saving devices to give more work to those who invest in them and to those who manufacture and distribute them? Is not the purpose to transfer work from poor people to rich people?

We say we want to help people in poor countries, but we often harm them. Perhaps the question we ask, "How can we help them?" is too difficult. In a situation like this I think we are well advised to take a page from the books of the mathematicians and in particular, to follow the advice of a mathematician named Jacobi, who said, "When you are up against an insoluble problem, invert it. Invert the question and try to answer the invert; then go back to the original question. Maybe you can then see the answer you really want."

I take for granted that we really want to help poor countries. I think there are few people who would say otherwise. But we've had little success in helping them despite great efforts. Maybe we should perform a Jacobian inversion and ask, "How can we harm the people in a poor country?" Let's try that question first. Afterwards we can return to the original question, the one in which we are interested.

So, how can we harm the people in a poor country? The answer is obvious, I think, to anyone who visits a poor country and notices three things. First, the widespread malnutrition; second, the chronic poor health; and third, the extensive unemployment.

In a typical poor country the unemployment rate is of the order of thirty percent, but this is a very conservative estimate. The various forms of hidden unemployment mean that a truer figure would be fifty percent or more. What are the employed doing? Nothing most of the time, because malnutrition and poor health sap their strength, making it impossible for them to exhibit the spontaneous high degree of activity that should be characteristic of any healthy animal, including the human being. If our intervention in poor countries were consciously guided by a malevolent desire to harm the people as much as possible, our course of action would be quite clear. We should attempt instantly to cure all diseases, eliminate all malnutrition, and reduce employment by bringing in labor-saving devices. The instantaneous conversion of fifty percent of the adult labor force into healthy, well-fed people with nothing useful to do, would, in the strictest sense, be revolutionary. Now, of course, you might say, "What they really need is a revolution, isn't it?" Perhaps so. But the question is, can people on the outside catalyze a revolution that will do any good, or even a revolution that will go the way the outsiders wanted it to go? Experience indicates that outsiders are not very successful at this. We don't even know whether we should hope that outsiders were more successful, considering some of the ill-thought-out dreams they have had in the past. Since simultaneously and instantaneously reducing disease, malnutrition, and work is the answer to the Jacobian inversion, it follows that the answer to the major question (the one we are really interested in) must differ with respect to at least one of the variables. Humanely we want to reduce malnutrition and disease; that leaves only one thing that we can seriously consider not reducing, namely, labor. (Maybe we should even try to increase it.) Paradoxically, a so-called labor-saving device is tolerable only if it is one which, in fact, increases the total employment (perhaps through causing employment in other jobs). I think this important principle should be our guide.

It is impossible to keep a healthy, well-fed animal inactive. Millions of years have selected for spontaneous activity and we are the result. We cannot normally remain inactive for long periods of time. Our grandfathers used to say "idle hands are the devil's workshop." We say it is condescending to say something like that now. But at the upper levels we've replaced the old saying with Parkinson's Law, which says "Work expands to fill the time made available to it." It's the same thing, just different words. Work is a funny thing, you know. There's no definition of it that fits all cases, so I think we'd better substitute the word "activity" for "work," and distinguish the kinds of activity. Parkinson's Law should really say, "Activity expands to fill the time available to it."

Activity must be paid for by calories of energy expended. Food calories can be divided into two classes. The total calories are composed of maintenance calories and activity calories. Maintenance calories are required just to keep our metabolism at the lowest level compatible with life. With maintenance calories we are alive, but we don't accomplish anything either good or bad. Activity calories are used for paying for everything over and above and demands of basal metabolism.

We can divide activity calories into two more classes: calories used for scheduled activity and calories used for unscheduled activities. Scheduled activity is roughly what we call work. Unscheduled activity is a class without limits. It is variously distributed among the arts, athletics, play, hobbies, exploration, random activity, riots, and revolution. Calories in must equal calories out. If people are fed more, their activity necessarily increases. This is certainly true at the lower levels of nutrition, which is our principal concern. If the increase in calories put in is not absorbed by what we call work (scheduled activity), we can be sure that it will burst out in unscheduled activity, which may be either creative or destructive of the proper goals of society. Does this mean then that we should not risk improving the health and nutrition of a nation or decreasing the workload? Not necessarily. It merely means that any conscientious interventionist will not try to improve health or nutrition unless he is convinced that at the same time he can find a useful outlet for the human energy released. This is not an easy standard to meet.

I think we had better look a little closer at what we call "work." What are its characteristics? We don't understand a great deal about it, but I think there are four things we can say about it. First, it is social. Second, it is rhythmic. Third, it is felt by the worker to be worthwhile. And fourth—and here I come to the only quantitative statement I'm going to make—it does not exceed twenty hours per worker per week on the average. Admittedly, the Thoreaus and the Einsteins of the world don't fit this generalization, but most people do.

About twenty hours a week is the normal workload of a human being. This is one of the surprising discoveries of recent anthropology. The twenty-hour-week seems to be universal among hunters and gatherers. By any reasonable definition of humanity, we have been hunters and gatherers for at least ninety-nine percent of human evolution, and something else for the one percent of evolutionary time. I think there's quite a bit of evidence that shows that we have not escaped from our evolutionary past. I can see no good reason why we should. Unless a fundamental reason can be found for escaping from our past, we should reject as abnormal any schedule that requires more than twenty hours of work per week for the average worker. The scheduled activities that we call work nominally demand forty hours of work per week in the society in which we find ourselves. But if you observe closely, I think you'll find that very few so-called "workers" violate the twenty-hour rule. Out of a conscientious regard for the laws of nature, workers institute coffee breaks (which get longer and longer), and impromptu conferences (which occur more and more). Committee meetings are the great escape hatch of university faculties. If you're in a business you run around checking up on what the competitors are doing, getting out of scheduled activity that way. What with one thing and another, through myriads of activities we officially call work, we manage to avoid scheduled activity about half the time. Thus do we obey our true nature. We are still hunters and gatherers at heart. Unscheduled activities, at their best, reintroduce play—but we must never call it play!

A hypothetical benevolent dictator, fearful of revolution, may make work in order to keep his people happy. His success depends on keeping his intentions secret. So what does he do? For one thing he might build pyramids. But pyramid-building, to be successful, must be justified by a transcendental myth. Our late, great, space program failed in this regard. Hence, it's now a dead duck. Because democracy is committed to the free flow of information, democratic governments are at a great disadvantage in making work and getting away with it. The WPA and the CCC of Depression days were speedily labeled leaf-raking, and the scheduled activity soon lost the legitimacy required in successful and unquestioned work.

The risk incurred in saving work in a poor country can be diminished in various ways: one by pyramid building; two by creating or continuing the tradition of turning out artistic handwork for export trade; or three by raising expectations with regard to the acceptable standard of life. The third way, raising expectations, is, of course, capable of great, though not infinitely great, expansion. Unfortunately, environmental impacts limit the possibilities of following this route. Psychologically, raising expectations also incurs the danger of fostering envy—which has dangers of its own. The ambivalent moral effect of the expectation-envy couple creates a serious ethical problem for any well-meaning interventionist.

In spite of all these dangers, I think that one of the great externalities of the Schumacher program of creating simple machines that require a minimum of capital in proportion to the amount of human effort put into them, is one of the most hopeful things that I have seen in my lifetime for helping other countries. This intermediate technology approach allows people to work their way out of their own troubles with a minimum of danger of creating the explosive power found in a large, well-fed population without sufficient employment.

Intervention along the lines laid out by Schumacher can keep people gainfully employed in improving their own situation. Changes that result from such activity are evolutionary rather than revolutionary. Achieving such change builds up the self-esteem of the achievers, thus leading to still more achievement in the best of psychological climates. People of good will cannot ask for more.

Economic Growth versus Human Development: A Report on India

P. K. Mehta

The information presented here is limited to India, but the general conclusions are also applicable to other countries of the Third World.

ECONOMIC GROWTH

About thirty years ago India won political independence from British colonial rule. Since then, the country has made considerable economic progress. For instance, recent statistics published by the Indian Government show that during the last ten years the gross national product increased from 27 billion to 72 billion United States dollars. The power generation capacity increased from 10 million to 21 million KW. There was a fivefold increase in the number of tractors (from 50,000 to 250,000), a twelvefold increase in the number of irrigation wells (from 93,000 to 1.13 million), and a fortyfold increase in the consumption of chemical fertilizers (from 0.8 to 32 million tons).*

The overall index of industrial production has almost doubled during the last fifteen years. The food grain production for 1976, which was only 51 million tons in 1951, is estimated to be 118 million tons. During this twenty-five-year period, the population increased from 361 million to 610 million. Thus, while the population registered a growth of 66 percent, the food production increased by 130 percent. Since 1950, the food grain production has

*"India Fact Sheet," Government of India, Directorate in Advertising and Visual Publicity, 1976.

risen by 2.8 percent per year, whereas the annual population growth rate was 2.1 percent. Mellor reports that the rate of food production per year in India outstripped the population growth by a considerable margin.*

POVERTY IN INDIA

A country, however, is more than just a piece of real estate. It has people, too. After all, what good is economic growth if it does not promote human welfare and human development? This question arose in my mind when, in the summer of 1975, I spent two months in India and was surprised to find little or no relevance of the aggregate national economic growth to the general human condition in most parts of the country. For example, in a small village in the state of Madhya Pradesh, I watched landless peasants transplanting rice. They were paid only two pounds of grain for ten to twelve hours of hard work. This was inadequate to provide two meals a day to all the members of a typical family. In order to survive they were forced to eat bread made from a wild flower (mahua flower). Moreover, since the farming operations were seasonal, there was often no work for them in the village. They lived in mud huts with straw-thatched roofs which leaked badly during the rainy season. There was no clean drinking water in the village, no sanitation facilities, no electricity, and no hospital. A dilapidated building, consisting of two small rooms, housed the primary school where children of grades one to five were huddled together under the supervision of one schoolteacher. Many children did not make it to school because they were needed to earn extra income for the family. Was the quality of human life in this small village an exception, rather than the rule? If the latter, how did the economic prosperity of the country bypass the rural poor? Let us now examine these two questions.

About 500 million of India's 610 million people live in half a million villages. One-half of the 70 million farm holdings are less than one hectare and, therefore, are unable to provide adequate sustenance to their owners. It is the poor farmers and their families, plus the landless who have sold or mortgaged their holdings to

---

*J. W. Mellor, "The Agriculture of India," *Scientific American,* vol. 235, no. 3, September 1976, pp. 154–163.

landlords, who make up the bulk of the 240 million people who are acknowledged by the Indian Government to be below the poverty line (50 United States dollars per capita yearly income). Since there were only 120 million people (33 percent of 361 million) below the poverty line in 1951, it is shocking that the number of poor in India has doubled during the last twenty-five years of self-rule.

Again, the statistics on poverty do not tell the full story, because the definition of poverty differs from one country to another. Poverty in India is a wretched human condition in which an individual's life, liberty, and happiness are constantly threatened. Poverty means stunted bodies and wrinkled faces of the young for lack of food; poverty means women who are unable to hide their shame for want of clothes; poverty means inhuman living quarters full of squalor and ugliness.

DISMAL PROGRESS IN HUMAN WELFARE

Evidently, the national developmental strategies failed to promote minimum human welfare, which is a necessary prerequisite for human development. India's progress report in the areas of health, housing, education, employment, and human rights represents the dark side of the coin of development. During the last decade, the number of hospitals rose from 14,213 to 15,235—an increase of 1,022 for a population increase of 100 million (one hospital added for every 100,000 people). The number of hospital beds was increased from 288,000 in 1965 to 388,000 in 1975 (one bed added for every 1000 persons). In 1961, there were 57 million rural and 11 million urban dwellings. By 1971, the number increased to 66 million and 16 million, respectively (14 dwellings added for every 100 people). In the last ten years, in spite of an overwhelming majority of the population being illiterate, the literacy rate rose by only 6 percent.

That the policy of industrial growth, and not human welfare, as an end in itself, is still being pursued vigorously by the government is evident from the revised budget for the current five-year plan (1974–79). The overall budget for health, education, and welfare has been reduced from the Draft Plan from 5.2 to 3.85 billion dollars—a reduction of 25 percent, which includes a 45 percent reduction in the outlay for primary education. The outlay for atomic energy is up from 12 million to 185 million dollars. The outlay for agriculture has been curtailed in spite of the fact that most Indians live in rural India and earn their living from agriculture. In December, 1976, addressing a Conference of Industry Ministers from twenty-two developing countries, India's minister of industry advised:

> No doubt agriculture should be transformed and modernized, but it was a well-known fact that the growth of productivity of labor was much faster in industry than in agriculture. Hence more rapid industrialization would lead to a more rapid rate of economic growth than would otherwise be possible for developing countries.

Is this not a case of the blind leading the blind?

The large majority of the population which exists outside the market economy is being bypassed when the national developmental strategies are guided by the Western and the Russian models of economics. For instance, during the last two years, the Indian garment industry increased its share of exports from 13.7 million to 262.4 million dollars per year. As a result, the availability of cloth has become scarce; and, due to export subsidizing, its price has become higher in the domestic market. While the foreign markets proudly exhibit the India-made cloth, millions in India cannot afford a change of dress. In fact, half-naked children, and even men and women, are a common sight in city slums and villages.

Similarly, India produces steel at $200 per ton, but subsidizes it for export at $150 to $170 per ton.* It is estimated that subsidies on steel export in the years 1976 and 1977 will amount to about 90 million dollars. The domestic price of steel was raised thrice, which caused recession and layoffs in the steel industry. Is this not a shocking example of the transfer of valuable resources in the reverse direction by a national government?

UNEMPLOYMENT IN INDIA

Again, escalation in industrial and agricultural production has been achieved without regard to, and probably at the cost of, another important human goal, namely that of providing productive work for everyone. Human beings can seek fulfillment in life only when

*B. M., "Lessons of SAIL," *Economic and Political Weekly* (Bombay, India), October 16, 1976.

they have an opportunity to use both their minds and bodies for productive and creative work.

The Indian Planning Commission's estimate of the total number of unemployed in the country today is about 21 million. The Central Employment Directorate reported that the unemployed and the underemployed will reach the astronomical number of 60 million and 30 million, respectively, by 1978. The unemployed include a large number of college graduates. According to a survey by Sharma and Apte in 1971, there were 288,487 unemployed college graduates including technical personnel, as compared with only 28,000 ten years before.* If the current pace of economic growth continues at an optimistic rate of 5.5 percent, the number of unemployed college graduates in the next fifteen years is expected to increase to 3 million. A similar forecast by Makhijani and Poole shows that during the next twenty-five years the total labor force in India will grow from the present 225 million to about 400 million, whereas a 7 to 8 percent rate of industrial growth will increase the industrial employment by 5 percent per year, thus creating only 25 million additional jobs.† Clearly, just to maintain the status quo in the employment situation, not to speak of finding productive employment for the already unemployed or underemployed, about 150 million jobs will have to be found within the agricultural sector of the economy.

Lack of food, housing, health care, and jobs is continuing to push many Indians into a state of destitution. A UNI report in 1973 estimated 5.5 million beggars in the country as compared with one million in 1961. Among the total beggars in 1973, 52 percent took to begging because of various diseases, 14 percent because of physical handicaps, and the rest for want of work. Describing the beggars living on the pavement near Delhi Railway Station Overbridge, Rajinder Puri writes:

> In the dark of the night, the stench of urine from the walls against which they lie huddled, and the rags covering them, discourage investigation. The whispering sound, as shy as a mouse in flight, is all that escapes from the beggars' world to

*G. D. Sharma and M. D. Apte, "Graduate Unemployment in India," *Economic and Political Weekly*, June 19, 1976.

†A. Makhijani and A. Poole, *Energy and Agriculture in the Third World*, New York: Ballinger Publishing Co., 1975, p. 5.

> give proof that the hunks of flesh and bone with faces frozen into masks and with stumps for arms and legs are human—thereby meaning that they too can laugh and weep and hope and despair.*

HUMAN RIGHTS AND HUMAN DEVELOPMENT

It is obvious that in India the first victim of the fetish of measuring the national progress by industrial growth, volume of exports, and GNP is human welfare. The second and more serious victim is basic human rights. For when the people agitated against soaring prices, unemployment, and rampant corruption, the ruling government extended its life by taking away the basic human rights of the people. In the name of national security and national growth, the Indian government imprisoned thousands of political opponents and curtailed the freedom of the press and the judiciary through illegitimate means, including fundamental changes in the Constitution by a Parliament whose mandate had expired.

In their preoccupation with the Western models of achieving faster economic growth, the Indian leaders lost sight of the direction of growth. As a result, the overall goal of human existence—human development—has been entirely forgotten. Nobody can be against growth, for growth is an essential feature of life. Every nation has a right to modernize and progress, but the key issues are what constitutes modernization or progress and what should be the direction of progress. Also, some related issues are whether an individual or a broad consensus should set the national goals, how the country shall be governed, how the leaders are to be held accountable for political and economic misdeeds, and by what process transfer of political power should take place.

INDIAN SOCIETY TODAY

Ten percent of the population includes landlords and rich farmers, politicians, big businessmen, officers in civil and defense services, lawyers, doctors, technical personnel, and industrial managers. Materialistically, they form the top ten percent, but morally and spiritually they are the bottom ten percent. Too few of them are

*Rajinder Puri, *The Wasted Years*. New Delhi: Chetana Publishers, 1975.

interested in establishing a just social order. Although they are in a political and economic condition to optimize their own human potential* and to accelerate the processes of human development in the rest of the society, generally their private life is devoted entirely to advancement of personal economic interests while cleverly maintaining a public posture of patriotism, community, and national interest. It is with this segment of society in mind that Rajinder Puri wrote:

> The Indian elite has corrupted the public to make the average Indian a most unpleasant, dishonorable, and demoralized individual. . . . There is nothing as contemptible as the ruling class Indian—who fawns upon the white man, apes him, and takes pride in accepting guidance from him.†

The other ninety percent of the society includes the destitutes, the landless, the poor farmers, the petty office clerks, the small shopkeepers, the industrial workers, and so on, who often live under inhuman conditions and find it difficult to keep the body and soul together. They are politically and economically at the bottom, but ethically superior, and culturally still deep rooted in the traditional humanistic values of the Indian culture. India cannot and will not grow materially and spiritually in the image of Tagore and Mahatma Gandhi until these people seize political power and transform the existing political, economic, social, and educational structure to make it possible for all members of the society to be productive and enjoy the fruits of such productivity.

SOLUTIONS FOR ABOLISHING POVERTY

From time to time, statesmen and scholars all over the world have offered solutions to the problem of human misery and degradation in the countries of the Third World. Many of these solutions are based on myths which must be destroyed, because they do considerable harm by misleading people and raising their hopes falsely. Foreign aid for rapid economic development, authoritarian systems of government, compulsory sterilization, and adoption of intermediate technology are examples of such solutions which tend to

*The Gandhian ideal of human development does not stop at material growth but extends into the realm of morals, ethics, and the spirit.

†Puri, *Wasted Years.*

divert public attention from the root cause of the problem. Let me elaborate further.

India continues to be one of the largest recipients of financial assistance in the form of loans and aid from the United States, Russia, European countries, and international agencies. This financial assistance has been a significant factor in India's industrial and economic growth. However, as already shown, the benefits from the economic growth have failed to permeate the entire society. In fact, a large portion of the financial gain has ended up in the form of unearned incomes for the politicians, top bureaucrats, and big businessmen, which widened the disparity between the rich and the poor. Would more aid not aggravate the disease rather than cure it? Can't we draw lessons from other countries where development was not limited by financial resources? Between 1960 and 1970 in Brazil—a relatively prosperous country—it was found that despite the six percent yearly growth in GNP made possible by the injection of large sums of money into the economy, the number of low-income people (bottom eighty percent) increased from 50 million to 65 million. In this context, let me quote one of the conclusions from a recently released six-year study by a team of twenty Latin American experts:

> International aid *in the conditions currently prevailing* in most developing countries would only contribute to increased spending by the privileged sectors and would have little or no effect on the living conditions of the majority of the population.*

Another myth which has recently deluded the public both in developed and underdeveloped countries is that a strong and centralized authority with dictatorial powers is better suited than a democracy for accelerating economic development and implementing radical measures aimed at removal of poverty. Again the lessons from other countries are instructive. In Mexico the recent disturbances over distribution of farm land, drastic devaluation of the Mexican currency, and growing disparity between the rich and the poor have dramatized that sixty-six years of one-party rule have failed to bring social justice to the people. Five years of martial law in the Philippines have increased both the ostentatious wealth of a

---

*A. Herrera, "Catastrophe of New Society," 1976.

few and the poverty of many Filipinos. Since the political and economic power remained in the hands of a coalition of factions comprising close associates of the president and the neutral military and technocrats, the change from old democracy to new authoritarianism did not change anything for the poor. Military rule in Pakistan ended by losing half the nation. Spain, Portugal, and Greece are struggling to recover from the effects of long spells of dictatorial rule. Chile, Iran, Thailand, Uganda, and many other countries are still under dictatorship. Whimsical policies, unaccountability of the decision makers, and ruthless repression of basic human rights are hallmarks of these authoritarian regimes, which are under no compulsion whatsoever to perform. Therefore, without democratization and decentralization of the political structure in the Third World countries, the poor will have no power to free themselves from misery—which is an essential prerequisite for realization of human potential.

Population control through compulsory sterilization is being suggested as another panacea for all ills of the Third World countries. The proponents of this idea claim that overpopulation breeds poverty by eating up the fruits of economic development and, consequently, minimizing the saving and investment potential. Barry Commoner argues that rather than overpopulation breeding poverty, it is the other way around.* A low standard of living is associated with high infant mortality, which along with a poor family's need to have more wage-earning members, causes overpopulation. As the standard of living rises, the infant mortality rate drops, and the population growth begins to level off. Both Commoner and Pathy contend that the population growth in the countries of the Third World is a result of the colonial exploitation (wealth produced in the colony was largely diverted to the advanced nation where it helped to raise the standard of living and lower the population growth), which is still continuing in the form of economic imperialism. Pathy argues that many Third World countries are rich enough in natural resources to support even higher population densities.† Also, since many nations registered a higher rate of

*Barry Commoner, "How Poverty Breeds Overpopulation," *Ramparts*. August–September, 1975.

†J. Pathy, "Population and Development," *Economic and Political Weekly*. July 24, 1976.

growth of GNP than of population, the real causes of increased poverty are concluded to be imperialistic domination and *internal distributive relations.*

The misconception of the population problem in the Third World by the Western leaders, and the mental slavery of India's ruling class to the West, culminated in a ruthless and inhuman campaign of forced sterilization of seven million poor during 1976. It was reminiscent of Hitler's treatment of the Jews. This is not to belittle the significance of the population problem but to point out that population is not the main cause of increased poverty, and that a more humane approach, such as better distribution of income and education in family planning, would put the burden of remedying a problem created by others on the voluntary participation of the individual victims of the problem. Commoner concludes by saying:

> If the root cause of the world population crisis is poverty, then to end it we must abolish poverty. And if the cause of poverty is the grossly unequal distribution of the world's wealth, then to end poverty *we must arrange to redistribute the wealth among nations and within them.*

Finally, let us examine whether or not adoption of intermediate technology can solve the problem of poverty in the Third World. It is gratifying to note that proponents of intermediate technology, such as Dr. E. F. Schumacher, have made the Buddhist and the Gandhian economics respectable in the eyes of the Western economists and their Indian counterparts. We are reminded that more and bigger is not necessarily better, and that rampant industrial growth of capital-intensive industries has brought about the crises of unemployment, pollution, inflation, and social alienation. Since intermediate technology is labor intensive, it is suggested that adoption of this type of technology for increasing industrial and agricultural productivity will distribute the fruits of productivity more evenly in the society, and thus promote social justice. This is no doubt a step in the right direction, but like the solutions discussed earlier, this will not help alleviate poverty unless the relationship of the economic dependence of the poor and domination by the rich is ended. For instance, only rich landlords have the money to set up rural industries; hence, by virtue of ownership of these industries, they will continue to exploit the poor. Inamdar points out that a product of intermediate technology, the Gobargas plant, was promoted only a few years ago by the government of India. Today, more than two-thirds of these plants, which produce gaseous fuel from animal waste, are owned by big farmers. He states:

> Appropriate technology does have its nonalienating aspects: it is indigenous, it is low cost, and it provides for utilization of waste materials. But no matter how far this policy of rural industrialization is carried in practice, it provides no solution to ending the *exploitive relationship under present forms of social control*. . . . Industrializing the rural areas is more a social and political task than merely a technological one.*

The passages underlined above show that it is the present political-economic structure which is the most important obstacle in the way of abolishing poverty in the Indian society.

CONCLUSION

The development programs implemented in India during the three decades of self-rule have brought about substantial economic growth. In spite of this, the number of people who have fallen into a dehumanizing and degrading state of poverty has more than doubled. This happened because the ruling elites, who inherited an expoitative political-economic structure from the British, have preserved the system to serve the interests of this small group. As a result, the Indian society today consists of two types of people—a minority which is materialistically better off, but morally bankrupt, and a majority which is suffering the indignities of poverty, but still retains the traditional values of the Indian culture.

Adoption of intermediate technology and family-planning programs cannot solve the problem of poverty unless the social, political, and economic institutions are restructured so as to become more responsive to human welfare. Authoritarianism in a pluralistic society such as India cannot bring about human welfare, and will thus hinder human development, which should be the national goal.

Human condition, and not mere statistics, should be the yardstick for measuring national progress. That which is helpful for human

---

*Z. E. Inamdar, "Ideology of Industrialization," *Economic and Political Weekly*. November 13, 1976.

development ought to be visibly growing in the society, and that which is bad for human development ought to be diminishing. And this will not happen until the present power structure, which has a slavish adherence to foreign norms of progress, is made to dismount from the backs of the oppressed majority.

Discussion

AUDIENCE PARTICIPANT: A question for Dr. McCalla. My question is, how can we expect these people to operate these intermediate or moderate technologies if the illiteracy rate, especially in rural areas, is often 95 percent? How can we train them to do that?

ALEX MCCALLA: I think my comments were in relation to alternatives which would be high-capital-investment, technical technology. That this was certainly a preferable way to go. I think Schumacher makes the point that there is a totality, a matrix within which that has to operate, which clearly involves the mechanism of education and training. But I think the other point he makes is also very relevant and that is that by certain kinds of standards we would define people as being illiterate, it doesn't mean they lack the capacity to do useful and other kinds of important things. And I think the entrepreneurial dimension does not necessarily have to occur, although it would be preferable, within the context of formal education. But I didn't mean to underplay the educational dimension. I think we tend to look at a development process all too often in a sort of sequence of bottlenecks and therefore we deal with one first and then come to another. Quite frequently we get the education first, but I think you have to deal with a set of things simultaneously.

AUDIENCE PARTICIPANT: Dr. Hardin, I would like you to comment on the policy of triage and its relevancy to your opinions.

GARRETT HARDIN: The policy of triage is a very simple one. It means that if you do not have enough goods to go around, it's better to focus them on the recipients where what you give them is most likely to do good. Now, I would imagine by the question you mean a broader question as to what my position is on helping other countries. I'm not against helping the poor, but I am very dubious about intervening in the lives of other people because my intervention may be help or it may not be help. This is what worries me. I think our record of intervention is quite poor. So my attitude would be this. To the extent that we have the ability and to the extent that other people are willing to let us intervene in their lives, let us do so, but very cautiously, recognizing that we may do more harm than good. Let's be properly humble about our ability to intervene properly. For one thing, I don't think we can save the entire world, just to put it bluntly. I think we should give our best effort to just a few cases where the opportunities are best, where the chances of succeeding are best. That's my basic position.

AUDIENCE PARTICIPANT: I'd like to direct this to Garrett Hardin. How can you say, in light of the studies that have been done on our aid to Third World countries which show such things as the rebuilding of the tiger cages and such things that our aid is not intended to be malevolent?

GARRETT HARDIN: Well, I'll grant anything that you want in this regard, because I don't want to be diverted from the important thing. Perhaps some of our aid is malevolent. If it is, I disapprove of it. What I am saying is, even the aid that is not malevolent, in which we intend good, that even this is really dangerous. And that's what I'm really talking about. You can have your malevolent aid and take it and do whatever you want with it.

AUDIENCE PARTICIPANT: My question is to Professor van Slyke. Do you see the difference in access to land and resources as between China and other developing countries as significant? I might include the United States too, for the poor here who do not have access to land and resources.

LYMAN VAN SLYKE: I do see access to land, resources and education in China as different from the picture that Professor Mehta has painted for India. I think this is a very substantial difference. Within China, however, there is differential access to these resources as well. There are some areas which are much better endowed by nature, by population, and by previous history with the capacity to advance more rapidly. These are in general the provinces and cities and their hinterlands in Eastern China. Many of the interior provinces are much poorer and one of the major issues within China is the degree to which richer areas should assist poorer areas. But overall, taking China as a whole, I do see the access to resources, to land, and to education, as much more equitable and much more generally available than in India.

AUDIENCE PARTICIPANT: So they don't have private owners standing there raising or levying private tax on the land before it can be used by someone?

LYMAN VAN SLYKE: Yes, there are no direct taxes because the land is owned collectively with the major exception of up to five percent of the land, which is retained in private ownership by peasants from which they are free to sell produce on the open market. But the bulk of agricultural land, except for the so-called private plot, is owned in common. Taxation is not upon the individual; it is collected by the state in the form of grain purchase requisitions from agricultural units and by purchaser-controlled prices of industrial output from those plants which are within the state plan.

AUDIENCE PARTICIPANT: I have a question for Professor Mehta. It seems to me that as an example of alternative technology in India, the work of Vinoba Bhave would be a good example. I just wonder if you have any comments on the work he's done there in voluntary redistribution of land and training the peasants in subsistence agriculture.

P. K. MEHTA: This is one of the followers of Gandhi who taught that by moral persuasion to the rich landlords, he could persuade them to part with some of the land which could be distributed to the landless. The movement was a partial success because of his personal prestige and powers of persuasion and because many other Gandhians joined it. They tried to show that there should be some part of morals and ethics in the society, that people trying to enjoy all the fruits of things by themselves is not fair enough. But on the other hand, most of the land they got was barren land and was not very good. And even when some land was given to the landless, they did not have the means, they did not have the tools of productivity like the oxen or the irrigation water, or the fertilizer. So it has to be a total restructuring. Many experiments like this have been done with partial success and blocks have been everywhere. Somewhere that system interferes.

AUDIENCE PARTICIPANT: Dr. van Slyke, in your listing of "the five smalls," I noticed you didn't include any of the subsistence consumer products, food, shelter, clothes. When one thinks of alternative technology one usually thinks of local decentralized production of subsistence goods. It seems to me that they've done so well that you didn't even call it industry anymore. I just wondered in your presentation why it ignored that whole segment of the industrial production?

LYMAN VAN SLYKE: Well, there's much more emphasis placed on small-scale industry in producer inputs of the type that I described. There is very little handicraft production of textiles for example; in China today, this is very limited. Most textile for personal clothing is bought, it's purchased. It's one of the two or three products that are rationed in China today to ensure general distribution. Some of the building supplies that are built in these industries, cement in particular, are used for human dwellings, not only for water conservancy works. In some areas we visited, surplus labor was used for quarrying rock for making dwellings out of cut stone. But the main thrust in in this area of producer inputs to support agriculture in particular, what the Chinese call farmland capital construction.

AUDIENCE PARTICIPANT: I have a question for Professor van Slyke in reference to the Chinese situation and taxation on grain production. Is that a taxation in kind, and also is it on the level of production or is it a taxation on the amount of land? What do you think that has to do with the production level?

LYMAN VAN SLYKE: Taxes are paid in kind rather than being converted to cash. It's tax on production, but it's not called a tax, it's called "state compulsory purchase." That is, the commune must deliver a certain amount of the principal crop, whether that be wheat, rice, or whatever it may be, to the state at a price set by the state. If it exceeds its grain quota, then it is given a bonus price if it wishes to sell that grain to the state. It can decide to retain that grain as common surplus to be stored in the commune, but it has the option of selling it to the state as a bonus. State compulsory purchase is sometimes forgiven or reduced if there are very difficult weather conditions, drought or flood or something of that nature. So, it is not an iron requisition in the sense that it cannot be forgiven or altered. But it is not computed on the basis of land area, it's computed on the basis of yield.

AUDIENCE PARTICIPANT: What effect do you think that has on the level of production where you are taxing how much they produce rather than on how much land they have to get as much production as you can?

LYMAN VAN SLYKE: Well obviously you want to produce as much as you can because as you raise output and as long as the requisition share remains constant, there is a larger surplus for reinvestment in the local unit.

AUDIENCE PARTICIPANT: This is for Dr. McCalla. One of the things that Dr. Schumacher says is that the developed countries can contribute to the developing countries by providing them with relevant knowledge. My question is, how do we determine what is relevant knowledge, on what basis do we determine relevancy? What would you consider appropriate aid from the United States?

ALEX MCCALLA: The question of availability of relevant knowledge as an input or as a mechanism of the system is, I think, relatively limited in a contextual sense and in an application sense. I think if one looks at the transferability of knowledge, one has to look at it in several levels or gradations. I presume if one discovers the 126th element, that's a transferable piece of knowledge that knows no national boundaries and knows no physical limitations as to its use. If one is talking about the adaptation or development of a variety of seed, the transfer of that knowledge in the form of a seed technology from the United States to Peru is not relevant in the vast majority of the cases. I think that in the longer run the question is in the importance to create the capacity to develop relevant knowledge in the developing countries. The sooner we can get away from the view that one can transfer relevant knowledge off the shelf, so to speak, from one environment to another, certainly in agriculture, I think the sooner we'll get at the question of dealing more specifically with the issues of location-specific generation and adoption of knowledge in that environment.

AUDIENCE PARTICIPANT: You are saying that we have to spend time in that country to learn how to develop it, the relevant knowledge?

ALEX MCCALLA: I think you have to, in whatever way you can, spend time in the developing nation to learn what is relevant, and that may be a limited ability. You can talk about hierarchical ways to do that in terms of assistance, either with training people from other countries here, which I think has decreasing value as far as agriculture is concerned, or in terms of providing assistance in the development of educational institutions and other mechanisms in those countries. There's no magical solution to it, I don't think.

AUDIENCE PARTICIPANT: This is addressed to anyone who can answer it. Since so many of the other countries suffer from balance of trade problems and from loan payment problems, how can we expect them to do anything but turn to high technology solutions to these problems?

GARRETT HARDIN: This is a Catch-22 type of question. We get an appeal from a poor country for help. There may first be a proposal to give them something, give them materials, give them things worth money, which means a gift of money. But then this is turned down because it's been said that this would hurt their feelings if they got a gift. In fact, they may tell us that too, they may say, "No. We want a loan." So, OK, we make them a loan. We make them a loan at interest, usually what's called concessionary rates, maybe half the international going rate in the international sphere, so we make them a loan. Then, as they fail to solve the problem they come back the next year for another loan, and the next year for another loan, and, of course, there are people in this country who stand to make money by these loans and they encourage this system too. We get a very nice system going which causes the debt to mount up high enough, and then finally what do you do? They've got so much debt they can't pay it no matter what. At that point, in the case of India, for example, when Moynihan was ambassador, we negotiated an agreement with India whereby we forgave three billion dollars' worth of loans. You see, in other words, retrospectively it was a gift after all. Well, who's exploiting whom? It's not clear. Whatever we're doing, it's not helping, for sure.

AUDIENCE PARTICIPANT: The point is that we are not helping them by doing this. We're locking them into a situation in which they don't have any other alternative if they are trying to balance their payments. If they are trying to pay back these loans, they are in a situation where they almost can't do anything.

GARRETT HARDIN: I don't think we have any disagreement. I think we agree that the intervention should be appropriate to their problems and not give them intervention that is inappropriate and manage to run them into debt in which we finally have to forgive the debt. This doesn't do anybody any good.

AUDIENCE PARTICIPANT: It seems to me that a lot of what we are talking about gets ultimately to the question of who has the power and what are their motives. Some people think we have to do away with private property. But I think that we have not placed sufficient limits on how much private property, or resources, can be accumulated. Would someone care to comment?

LYMAN VAN SLYKE: I don't see any operational way to go from here. I'll admit that some people have too much private property, but that's the other guy, not me. You know that's the way we all feel, so where does that get you?

P. K. MEHTA: I think basically what you are suggesting appears to be the real problem because, if you were to see rural India, maybe ninety percent of the good land is owned by very rich people and the poor are either landless or very small peasants. About fifty percent of the farms are less than one hectare or about two and a half acres. And if they have the land, they don't have the oxen or the plow or they don't have the irrigation water. In terms of how can you take advantage of such a big human potential, of such a human resource, in terms of giving them a productive work and a creative work, since they have two hands and two feet, they have minds, what do you do? They may not be educated, but they can think, they are human beings, so in that sense you have to provide them certain boons of production. Where do they come from? They cannot come from foreign aid without restructuring the system. It doesn't help. So there has to be some distribution of the private property. And it has to be linked with how the gains of productivity in a community are to be divided.

LYMAN VAN SLYKE: I think that it isn't only a question of private ownership but it is the use and control of property and power. This is one reason why I think some of the things we see in the Soviet Union and elsewhere, perhaps to even some extent in China, raise certain questions. It isn't private ownership but it is access, differential access to the use of property and power. And control may be distinguishable in some cases from ownership. More specifically with respect to China, the Chinese have had and are in the process of making their own choices. But it is not my purpose to comment on that. One of the things that has made China able to avoid some of the difficulties that I think India has encountered is to have reduced and distributed private property and its control to a very large degree. But also, corollary to that is a requirement of a much higher degree of control than many societies would permit. The Chinese peasant, the Chinese farmer, is essentially bound to the soil in China today. He is not permitted to move or migrate to the cities freely. If he does so, he is breaking the law and he is eligible to be sent back and perhaps to be punished in some fashion. Now that is a price that some societies would not be willing to pay. But one consequence of that is that Shanghai does not look like Calcutta or Bombay. It isn't heaven on earth, but by binding people to place, one achieves that. But societies have to decide what price they are willing to pay in order to get something else. There is, I think, in this business, no free lunch.

VI

CHANGING THE SCALE OF TECHNOLOGY: IMPLICATIONS FOR LABOR, MANAGEMENT, AND ECONOMIC GROWTH

John Kemper

Tom Bender

Herman Koenig

Louis Lundborg

E. F. Schumacher

Sim Van der Ryn

INTRODUCTION: *Richard C. Dorf*

MODERATOR: *John Kemper*

Introduction

Richard C. Dorf

Traditional economics has considered the economies of scale obtained when the size of an enterprise is increased. The economic advantages realized by large organizations have resulted in ever larger concentrations of capital, labor, and technology. Traditional economics has also called for the growth of the firm and organizations in order to remain competitive and efficient.

Schumacher has challenged the wisdom of the centralized, large-scale, capital-intensive economy. Rather, he calls for a small-scale, decentralized, economically sound, labor-intensive organization which is appropriate to the local setting. The scale of organization must be treated as an independent issue. Is a large organization with alienated workers and loss of a humanely reasonable dimension the most economic? The concern of a society is equally for the production of goods and the quality of life available to its citizens. People need a sense of place within the economic system as well as within their geographical area.

Schumacher experienced this issue of appropriate size while he was with the British Coal Board. The Coal Board is an enormous government-owned monopoly controlling the mining, transport, and distribution of coal within the United Kingdom. Under Schumacher's leadership, this large-scale organization was divided into a confederation of human-scale semi-autonomous units. Self-contained quasi-firms were established for seventeen separate mining regions in order to obtain appropriate, person-oriented units.

From a traditional economic point of view the typical response of a production unit is shown in figure 1. As the number of input units

FIGURE 1

OUTPUT PRODUCT AS A FUNCTION OF NUMBER OF INPUT UNITS

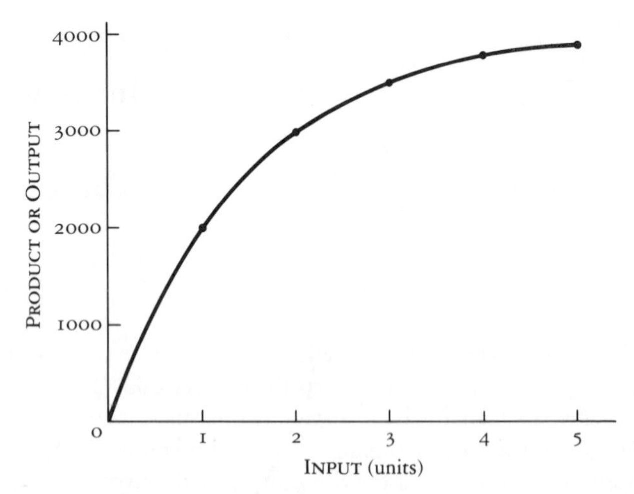

is increased, where the input may be labor or capital, for example, the output is increased. However, for each added unit of input the extra or added margin of output decreases. Thus, as shown in figure 1, and summarized in table 1, the marginal increase in

TABLE 1

PRODUCTION OF A HYPOTHETICAL PLANT
AS A FUNCTION OF NUMBER OF INPUT UNITS

Input Units	Output Products	Marginal Increase in Product
0	0	
1	2000	2000
2	3000	1000
3	3500	500
4	3800	300
5	3900	100

product is 2,000 for an increase of one unit of input between 0 and 1 units of input. However, the marginal increase in product is only 100 for an increase of one unit of input between 4 and 5 units of input. The law of diminishing returns states that the extra output resulting from the same additions of extra inputs becomes

FIGURE 2
ECONOMY OF SCALE

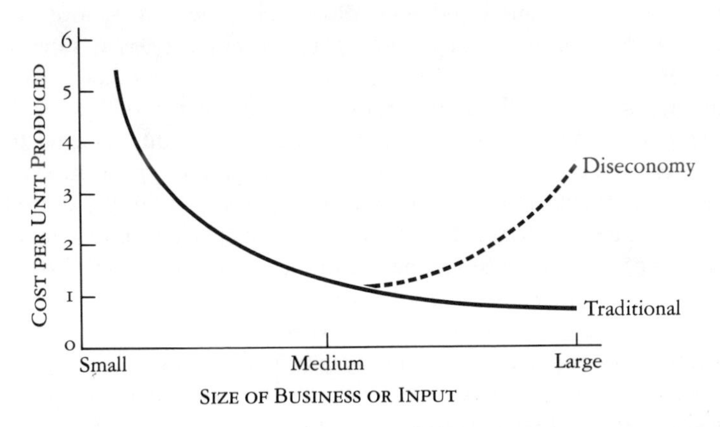

less and less.* The reduction in the cost per unit produced as a result of increased size of business or of an input such as labor is shown by the solid line in figure 2. Thus, it becomes more difficult to reduce the cost per unit as the size of the organization increases. Nevertheless, the so-called economy of scale does permit a large organization to reduce its cost per unit produced and remain competitive.

TABLE 2
CHANGES IN CALIFORNIA FARM SIZE AND INCOME PER ACRE

Year	Acres per Farm	Investment per Acre	Net Income per Acre (in 1959 dollars)
1950	260	$158	$23.00
1959	348	$377	$24.77
1969	617	$544	$21.58
1971	654	$533	$21.54

As an example, technology and scale have had a significant effect on California agriculture. As noted in table 2, farms have become larger, the investment per farm has increased, while the net income

*P. A. Samuelson, *Economics: An Introductory Analysis.* Eighth Edition, New York: McGraw-Hill Book Co., 1974.

per acre has decreased.* Economic conditions and technology have worked together to make farms larger and more dependent on investments of capital and technology. However, as is noted in table 2, the marginal net income per acre has decreased slightly between 1959 and 1969 (or in other words, the cost per unit output has increased a small amount between 1959 and 1969).

In many cases, organizations and production plants are said to experience a diseconomy of scale as the size of the organization or the input grows large. This type of effect is shown by the dotted line in figure 2, where, after a minimum cost per unit is obtained, the cost per unit increases as the size of the organization is increased. In a recent article, Koenig shows that as the capacity of a cattle feedlot increases beyond a minimum cost point, the cost per unit of output (dollars per gain of cattle weight) will increase.† This result also occurs when the size of a city exceeds a reasonable population or the size of a firm becomes unwieldy.

American industry and commerce includes both large and small firms and incorporates both highly centralized and widely decentralized companies. The largest United States firms, Exxon and General Motors, both had sales of about $48 billion in 1976. As reflected in the 500 largest firms, the typical big United States corporation had sales of $1.4 billion and showed a net profit of $72 million.‡ However, no single company accounted for as much as 4 percent of the total sales of the top 500. The top ten combined accounted for 22 percent of the total. The corporate power is dispersed over these 500 corporations as well as among thousands of smaller ones. One may conclude that the United States economy depends on large-sized corporations, but if their production and authority is decentralized and dispersed, a sense of belonging and place may be retained.

In the following articles the panelists reflected on the appropriate size of the firm and the scale of technology, the limits to growth, and pressure to achieve economic growth. Even if a common objective is agreed to, the question remains: How do we achieve it?

---

*B. B. Burlingame et al., "Technology and Scale," Agricultural Extension of California, 1972.

†H. E. Koenig et al., "Energy and Agriculture," Proceedings of the Workshop on Energy Extension Services, University of California, Berkeley, July, 1976.

‡"The Forbes 500's," *Forbes*. May 15, 1977, p. 156.

Changing the Scale of Technology

John Kemper

Yesterday the conference dealt with agriculture, the ethical basis of intermediate technology, and Third World development. Today we are to deal with technology itself. This morning with the question of scale, and this afternoon with energy and new technologies.

I noticed that the moderators yesterday did not make statements, but I would like to make a statement and, therefore, will make one along with the rest of the panel.

Let me start by saying that I do believe there are limits to growth and that we are now encountering some of them. My personal belief is that we are going to have to run as hard as we can just in order to stay in the same place, and even then we may not succeed in doing so. There is one thing I'm well certain of though. Whatever we think today are going to be our serious problems ten or twenty years from now, there's a high probability that we're going to be wrong in our predictions. To convince yourself of that you should think back to what predictions were being made ten or twenty years ago as to what our problems would be today and you will see what a dismal record we have about predicting the future. I think our problems will be serious ten or twenty years from now. Some of the things we are predicting today will probably be present, and I suspect strongly there will be some very serious problems in there that we haven't even thought of. I should call attention, though, to one individual who, twenty years ago, was predicting correctly one of our major problems today. This was M. K. Hubbert, a senior geologist for the United States Geological Survey, who was publishing the fact that fossil fuels were going to start diminishing just about now. He was substantially ignored at the time. But the fact is he was right. Today, of course, his views have been adopted pretty much as the conventional wisdom. But there have been many other people predicting things that didn't come to pass, and the trouble is to pick out the right ones from the wrong ones.

Now before we came to this conference—and certainly I had this view—we might have supposed that Dr. Schumacher's view would be that smallness is an essential ingredient of appropriateness. But that was before I read his book, and before yesterday. I would like to make a small quotation from Dr. Schumacher's book, which indicates that this isn't all of what he means. His argument is actually with the view, which is held by many in this country, that bigger is better. To quote him precisely, he says,

> For his different purposes man needs many different structures, both small and large ones. For constructive work the principal task is always the restoration of some kind of balance. Today we suffer from an almost universal idolatry of giantism. It is therefore necessary to insist on the virtues of smallness where this applies. If there were a prevailing idolatry of smallness irrespective of subject or purpose, one would have to try to exercise influence in the opposite direction.

Now I find this to make eminent good sense and I'm in instant accord with Dr. Schumacher. He would, I think, call it "finding the middle way."

Interestingly enough, yesterday almost every speaker readily agreed with that same point of view. I found with the earlier speakers there was no automatic assumption that smallness is the essential ingredient of appropriateness; I did not feel I ended the day in possession of a reliable guide to the manner in which any given speaker would be able to distinguish appropriate from inappropriate technology in advance of the fact. It's not too difficult to decide after it's all over whether you thought it was appropriate or inappropriate, but the big question is: How are we going to do this in advance? I admit that I'm not going to be able to do much better. It may even be something like sin. We all know what it is but we can't set down any guidelines for defining it. What we must do, I think, is in each instance to evaluate the presumed costs and the presumed benefits of a particular technology and to compare these against our own personal value systems. That's what we all do. One difficulty, though, is if each of us does this and with perhaps five hundred people in the room, you could have five hundred different conclusions as to appropriateness. I think the best way to communicate my own view of appropriateness to you is by some examples of the way I compare these against my own value scale. Let me give you some examples of what I consider to be inappropriate technologies.

My first and best example is the SST. It is not widely known that it was the accumulated wisdom of engineers and economists which played a large part in stopping the development of the SST. Some individuals were clamoring that the SST was a technological tour-de-force which the United States simply could not back away from—and furthermore, look at all those jobs it would create. There were others who pointed out that the SST represented a gigantic use of public funds for a purpose that could not be justified. It would be a huge user of natural resources—especially energy, and it could not produce an economic payback on any reasonably projected basis. Now it's easy to talk like this today because everybody knows the Concorde lost fifty million dollars last year. But my point is that many engineers and economists stood up at the time when it counted, and said the SST was not appropriate. They rejected the idea, advanced at the time, that we must build the SST in order to maintain our image of technological superiority, and they especially rejected the notion that the SST was desirable because it would be a gigantic make-work project to keep scientists and engineers from being unemployed. Clearly, they felt such an activity to be inappropriate because it was simply lacking in benefit, without it even being necessary to mention sonic boom noise or the ozone layer.

Space exploration—and my apologies go to my friends in the aerospace industry—is another matter whose appropriateness continuously needs to be examined. We all know why we got into the space race in the first place, but today it is not so clear as to why we are still there. We are told, of course, that we are acquiring priceless basic knowledge about our solar system and that we should not force questions of applications or value upon basic research. It's only the engineers who do that. Basic researchers don't buy the point of view that potential applications must be used to justify the value of basic research. Basic knowledge is valuable for its own sake. This is the kind of argument, of course, which university professors find compelling, and I'm a university professor. Professors are committed as an article of faith to the pursuit of basic truth and to the value of basic research, and I do reaffirm here my support of those concepts. And yet space exploration is enormously expensive and I would ask: Is it impossible to place limits upon costs and the use of resources while simultaneously asserting our support for the truth? I hope these are not incompatible, and so I would place space exploration on my list of items whose appropriateness should be continuously examined.

I also question the appropriateness of large solar electric power installations. I noticed that Mr. Van der Ryn's Office of Appropriate Technology does the same. Now I'm a strong solar enthusiast and I do expect solar power to become progressively more important to us in coming decades. But I also know that solar radiation is very diffuse and that we must necessarily do our harvesting of this energy resource over large areas of the earth's surface. I object to the notion of using up additional large undeveloped areas by covering square mile after square mile with solar collectors. Many people seem to think the desert is just fine for this purpose. To them, desert lands are wastelands. I do not agree and I do not think it is necessary to take this course. We've already covered large areas of the earth with buildings, homes, schools, stores, offices, warehouses, and factories, and I think we should utilize the roof areas of built-up places before we consider using up even more of the earth's surface, even for such a desirable purpose as solar energy.

One more example. Everybody likes hydroelectric power because it's clean and renewable, but not many people like large dams. Again, we have a question of appropriateness and we can't have it both ways. As a matter of fact, hydroelectric power may be one of the most dangerous forms of energy we possess. It's a matter of historical record that hundreds of times as many people have been killed from dam failures as by nuclear accidents. But while we are talking about such things, perhaps we should also remind ourselves that coal-fired power plants represent another dangerous form of energy, as any coal miner's family can tell you. During the decades from 1910 to 1940, United States miners died at the average rate of two thousand a year. Even in as late a year as 1970, 254 persons died in United States coal mines. The question of appropriateness is before us in its most human and terrible terms. In the face of such dreadful statistics it is possible that there might be some in this room who would state their willingness to forgo all forms of electrical energy from now on. But I doubt that there would be very many who would do that. We don't want to be reminded that people will be injured and some will die in the process of bringing us energy. Yet, it's a fact. And people are injured and killed in the process of bringing us food too. We can, and we must, do all we can to reduce the numbers of injuries and deaths, but we will never reduce them to zero.

Incidentally, perhaps it's worth pointing out that the great reduction in coal-mining deaths is at least partly because of the increased use of mechanization. Fewer men are sent into the mines; machines do all the work instead. Here we have a familiar two-edged sword. With fewer jobs, men are thrown out of work and that's one of the great problems in Appalachia today. But also if there were more jobs in the mines, then more people would be killed or debilitated by lung disease. One might say, "Why must men do such dangerous work at all since surface mining is much safer?" But then we have many others who strenuously oppose surface strip mining and urge us to do deep mining instead. And some of these may even propose that we just forget about our coal resources. Yet coal is our most plentiful energy resource. How do we sort out the appropriateness in all of this? And how is the question of smallness, for example, going to be very useful? I recall some small coal mines that I saw in West Virginia. They were backyard coal mines and they are the most dangerous of all.

While we are on the topic of safety, it has also been suggested that we should look seriously at the safety implications of solar energy. The second greatest cause of accidental deaths in this country is from falls, sixteen thousand deaths a year in the United States. Widespread use of solar energy will imply a vast increase in the amount of rooftop apparatus—apparatus that will have to be maintained and serviced much more frequently than the existing rooftop structures. Well, I'm a solar enthusiast and I suppose I'm going to have to be one of those people who will have to go up there and maintain the equipment. I don't want to fall. But on a statistical basis you can predict that there will be an increase in accidental falls. Well, there it is. We have to face these matters. Mainly my point is, there is no such thing as an absolutely safe technology or non-polluting technology. In this regard one of the most puzzling contradictions I observe in today's world is the insistence by some that we should burn more wood. Everybody loves the smell of a wood fire but if we all did it, we'd have an air pollution crisis of the first magnitude. Besides that, there are implications of deforestation on a vast scale. I would remind those who seriously suggest that we turn to wood in a major way that the reason Great Britain had to shift to coal in the seventeenth century is because they ran out of wood for fuel.

So what about some appropriate technologies, and I admit it's my personal value judgment at work and my values may not coincide with yours.

To begin with, when I come home at night and I flip the switch, I want the lights to go on. We probably could do with less electricity than we do and as a matter of fact, we are probably going to have to. And we could probably use less in ways that don't seriously affect our standard of living, but we do have a growing population, more and more people, whether we want them or not. They are going to want their share of services and they are going to want electrical energy. It's one of our most useful forms of energy. For the time being, the production of electrical energy is dependent on big power plants and intricate transmission systems. In my opinion, electrical energy is appropriate.

Another example is food. We dealt with it at great length yesterday. The greatest resource of the United States, in my opinion, is its ability to produce food. Maybe we can produce more of it than we have in the past. Many speakers yesterday seemed to think that we could and I hope they are right. But we are going to have a great increase in population coming at us, one hundred million more Americans in the next fifty years, even with the two-child family. Without an extensive technology in agriculture, I fear for our ability to produce food for these people, let alone to try to share some of it with other countries.

The wide availability of phonograph records and stereo systems is a great boon to the human race, and the result of a highly appropriate technology, in my opinion. I love music and I'll wager most of you do too. To me, it's almost like a miracle, the fact that we have this marvelous music available to us at a price we can afford. If I were to lose this particular boon, there's no doubt in my mind the quality of my life would have been degraded. This must be a widely shared value, because millions of records are sold every year. If we counted up the number of records owned by just the members of this audience, I suspect it would be in the tens of thousands. If you take five or six hundred people and you only assume each has twenty records or so, you've got ten thousand records. But those records are vinyl and vinyl comes from chemical plants. And chemical plants aren't small and they aren't non-polluting.

Well, I hesitate about mentioning this next case because it's going to be controversial to some of you. It has to do with my automobile. To be sure, my car is a Datsun compact. I get good mileage and I really see no reason why people generally in the United States can't get along with small cars. But why is that car so important for me? Not for commuting, because I ride a bicycle. I have the great privilege of living in a town like Davis where that is possible. But my car takes me to the mountains where I go camping, hiking, and backpacking, and these are pretty important activities to me. Now I suspect that many of you must share these same values because when I get to the mountains, most of the rest of you are already there! Now I would like to be sure that this particular bit of technology remains in place as long as it is possible and reasonable for it to do so.

I want to close on one point: energy conservation. I'm a great believer in conservation, but it does have many facets. Many persons would have us conserve by changing our life styles. We probably are going to have to change our life styles; but no more than is absolutely necessary, I hope. But I think we can make an enormous impact by simply using our energy better. It certainly is done in Western Europe where energy has been expensive for a long time. They have long since gone to things like co-generation of power, flash water heaters, small cars, things of that sort. More efficient systems of heating and cooling can work wonders for us. Spark-ignition pilot lights are already coming in. Heat pumps and fuel cells are penetrating the market and these have an inherently greater efficiency and less pollution than existing methods, especially when used with total energy systems. Much can be done. All of these meet my personal test for appropriateness because they supply our needs at the right time, in the right manner, and with greater effectiveness. None of them meet the test of smallness in the conventional sense, because they do rely on sophisticated technologies. But I think that's one of the things Dr. Schumacher is telling us—that growing automatically and unthinkingly always bigger is not in our best interest, and that extracting the maximum we possibly can from the resources we have is very much in our best interest.

Appropriate Technology

Tom Bender

It's easy for us to sit in a room like this and debate abstractly the possibilities of appropriate technology without making connection with our own lives and our situation here and now. We've been discussing whether more appropriate technologies can be developed, yet this room is full of people who already have done the things we're wondering about and there's no structure in this conference to allow each of us to make connection with the others and deal in practical realities instead of abstractions.

It's my feeling that institutional "technologies" like this are more at the heart of technological problems in the United States than the problems of machinery. Along with that, I feel we have been ignoring the potential for a *better* quality of life in adjusting to resource limits we are facing. Let's look at both of these issues in our real situation here. We've been sitting for two days in this air-conditioned, artificially lighted room, talking about energy conservation while it is beautiful, sunny, and 74° outside. Yet no one has even suggested that we do anything differently. This auditorium is actually well designed—all we have to do is pull back the curtains, open the doors, turn off the lights, and shut down the air conditioner—and we're in a beautifully naturally lighted space with soft, fragrant breezes instead of stale cigarette smoke. There—feel the change already?

When people began asking me for examples of appropriate technology, I very quickly found out I wasn't really interested in it. We already have countless such technologies all around us, but we often forget about them and move away from them. What really interested me was why we choose and have chosen technologies that under many situations are inappropriate. And the more I looked at it, the more I began to realize that the dreams we have of what we want to be and where we want to go are much more important than the technological skills of developing particular hardware. We have that. We have a lot of skills. We can develop what we need. What we really lack is a sense of where to go.

The dreams which have pretty much guided this society over the last hundred years or so were appropriate dreams and the technologies we developed were appropriate for the conditions we had of bountiful and seemingly unlimited resources. And it was pretty much in our self-interest to develop things the way we have. We know well the fate that has befallen other cultures which have not chosen the path we did during this period. They have been either physically or culturally decimated, as we ourselves narrowly avoided in two world wars. I don't think there is any reason to feel guilty about the choices we have made. We may wish now we hadn't made them, but I think generally there were a lot of reasons we did.

But obviously conditions are changing now. We're rapidly finding limits to the material growth we've pursued. And most important to me, this means totally different principles about how the world operates—new ways we must behave, new dreams that can be pursued, and different options that are open to us. We still seem unnecessarily afraid of these new things. Almost every report I've seen dealing with energy crises has said that we can reduce our rate of resource use by maybe ten or twenty percent, without any change in our quality of life. That phrase is so frequently repeated like a litany that it seems we're afraid to ask what's wrong with making a major change in the quality of our life. Could it be possible that a major change in life style could mean a better quality of life in the process of adapting to the situation which we are facing today?

That idea may be hard to grasp while we are still tied up in the growth world in which we've grown up, because an equilibrium society operates differently than a growth society. In fact, it's almost upside down. The rules that work in one are just the opposite of what works in the other. Basic things like the production of goods and services are unquestionably accepted during a period of growth, since they probably give us more reward than other ways we can use our energy and resources. But once we're in a situation where the resources are limited, where the best we can do is to sustain what we have available to us, things turn upside down and we begin to realize that the production and consumption of goods and services has become just a cost of staying even—a cost we should probably *minimize.*

Most of the structures we have built up in our society have been focused strongly toward the encouragement of growth, and quite reasonably so. Yet the changing situation suggests again that we should now turn many of these things upside down. I've been proposing recently that we eliminate advertising in public media. When the resources are no longer there to satisfy unnecessary desires which are generated by this kind of structure, when the centralization of market power which it causes is undesirable in terms of economic productivity, efficiency, and political stability, it's time to change our attitudes about advertising. It now seems wrong to use things like radio and television, which are shared public media, to create desires which we know we will be less and less able to satisfy.

Our dreams choose our technologies, and our dreams reflect the changing possibilities of our time and situation. Today a sense of a new dream is beginning to come into focus. All the way through architecture school we looked at pictures of beautiful buildings which were built in all periods of history. And it just flashed on me a few months ago that none of the Golden Ages of all those different cultures occurred during a period of growth. They did not occur during a period when everyone's energy was focused on expanding a culture physically or materially. They seem to have occurred when these limits had been reached—when people discovered that the expectations they had, in terms of personal fulfillment, well-being, beauty, or happiness were not attainable through these means. They seem to have suddenly done a hard accounting of where they'd been putting their resources and discovered, as I think we are discovering today, the incredible amount of resources that we put into stimulating a lot of these growth structures. Those resources, if put to other ends, can provide a wise society, a gentle society, a comfortable society, which has in the past, and I think even more today, the potential of being a better quality of life than what is attainable during a period of growth.

We always seem to focus on what gives us the best rewards or the best return for our own energy. Once we realize that the game we have been playing is no longer giving us as much return as other games, we move into something which is going to give us a lot more value. We can see that in the sudden fascination with gymnastics that developed from the last Olympics. It's beautiful!—people simply focusing on themselves and their own movement and grace and seeing this as a beautiful thing rather than requiring some mechanized intermediary through which to exercise their skill such as in auto racing.

Changes in our dreams, technology, and way of life begin to make visible some of the linkages normally invisible in everyday life, and give us opportunity to reflect on them and change them to respond to new senses of reality. We are rediscovering that which should be obvious—that values form the framework for what we call economics. Again I return to building as an example. Almost any practicing architect very quickly realizes that there are very severe economic restraints on building and that a lot of the beautiful fantasies they would love to build are limited because of desires to make the buildings economically profitable or economically efficient. If we step back and take a slightly broader look at the picture, we realize that our choice of economics has already been framed in by other choices. Before we figure the economics on buildings, we've already chosen to require certain structural safety standards. We have already chosen that we will design our buildings to be earthquake resistant rather than as cheap as possible. We've already chosen to design our buildings with special exit facilities in case of fire. We've already made a whole range of value-based decisions on what we feel to be desirable before we bring economics itself into play. It is within those frameworks that we then perform economic exercises which help us keep fairly tuned-in to not being excessively wasteful in the allocation of our resources. It is through frameworks like this that our new dreams—our discovering what we want to be and where we want go to—find concrete means of transforming our lives.

Questions were raised last night. What can we do? What is happening? Are things really changing? Are we dealing with these changes at all? For the last couple of years, we have been trying through *RAIN*, our journal of appropriate technology, to keep track of what is happening and trying to help it along. It seems that there *is* an incredible amount happening all over the country and all over the world in terms of developing good, satisfying, viable ways of doing things, and ways in which we can very readily join in. Our dreams *are* finding new ways of concretely transforming our lives.

We are discovering that a lot of the patterns in which we allocate our money and resources are perhaps not as wise as we would wish. There have been studies made which show that it is to the advantage of any community and even to banks in a community to ban franchises, to discourage outside industry and business from coming into a community, and to focus within a community for providing for its own needs. The Institute of Local Self-Reliance in Washington, D.C., did a very simple and effective analysis of the financial report of the McDonald's hamburger chain. They found that 65 to 85 percent of the total cash flow into a McDonald's hamburger stand leaves the community. That means lost jobs to the community. Even for a local bank it seems wiser to support development of local restaurants and food stands in which the money, the energy, the wealth, and the ability to deal with things remains and circulates within the community, strengthening the community rather than being exported to someone else.

The same thing has been realized about sewers. It now costs more than $4,000 per house to install a sewer system. We are told by authorities that there are no viable alternatives. But when you put a sewer into a community, you are being totally redundant, since everyone living there already has his own septic tank or sewage treatment facilities. People who move into a community later think that they don't have to pay for the sewer, that it's already there. But they don't realize that they are already paying *twice* for it—once in taxes and then in the cost they are paying for the lot or the house. Developers jack up the price of lots when a sewer is installed by what a septic system costs, because a lot which has a sewer available is equal in value to one which requires the installation of a septic tank. People have to pay such costs in their home mortgages, ending up paying a total of two of three times the original cost. Compared to the cost of alternatives such as composting toilets, you end up with a total cost for sewers of about $10,000 over twenty years—five times the cost of a septic system and as much as two hundred times as much as an owner-built composting toilet. Building excessive systems like these literally pours our wealth down the drain.

When we look like this at where we put our resources, we begin to realize that the appropriate technologies being developed today have much broader applicability than what we would think when we just look at them as a technological alternative. I think we are opening up new options for a better quality of life, and gaining a sense of what to choose, what way to go. We have the potentials for some really fine changes coming—not something we should be afraid of.

Appropriate Technology and Resources

Herman Koenig

The theme of this panel session is, can intermediate technology be brought into the mainstream of the United States economy, and what are the social, economic, and political implications of doing so? From the discussions we've had so far, I really don't know exactly what we mean by intermediate technology. We've referred to this on a scale of 1 to 12, and so it is very difficult for me to address the question as to whether intermediate technology itself will be brought into the mainstream of the United States economy. I've tried the term "appropriate technology." But I have difficulty with that also, because I have to ask, appropriate to what question? What I can say without qualification is that Americans will have no choice but to alter their life styles dramatically in the decades ahead. I say that not from an ideological point of view but from a very pragmatic understanding of the physical constraints of our changing resources base. Hopefully, many of these adjustments will have a salutary effect on our way of life—reduced environmental degradation, and increased quality of our social environment, and a more satisfying life.

I must also point out that my basic hypothesis is that a highly centralized, specialized, and very energy-intensive society is a product of the technological adaptations over a period of time in the history of the human species when energy has been at an extraordinarily low cost. I view technology not as an objective in itself but as a parameter of adaptation. The real fundamental issues are not technology. They are the economic, social, and cultural forces that have given direction to that adaptation. If this hypothesis is

FIGURE 1: DEPLETION HISTORY OF LAKE SUPERIOR DISTRICT IRON ORE

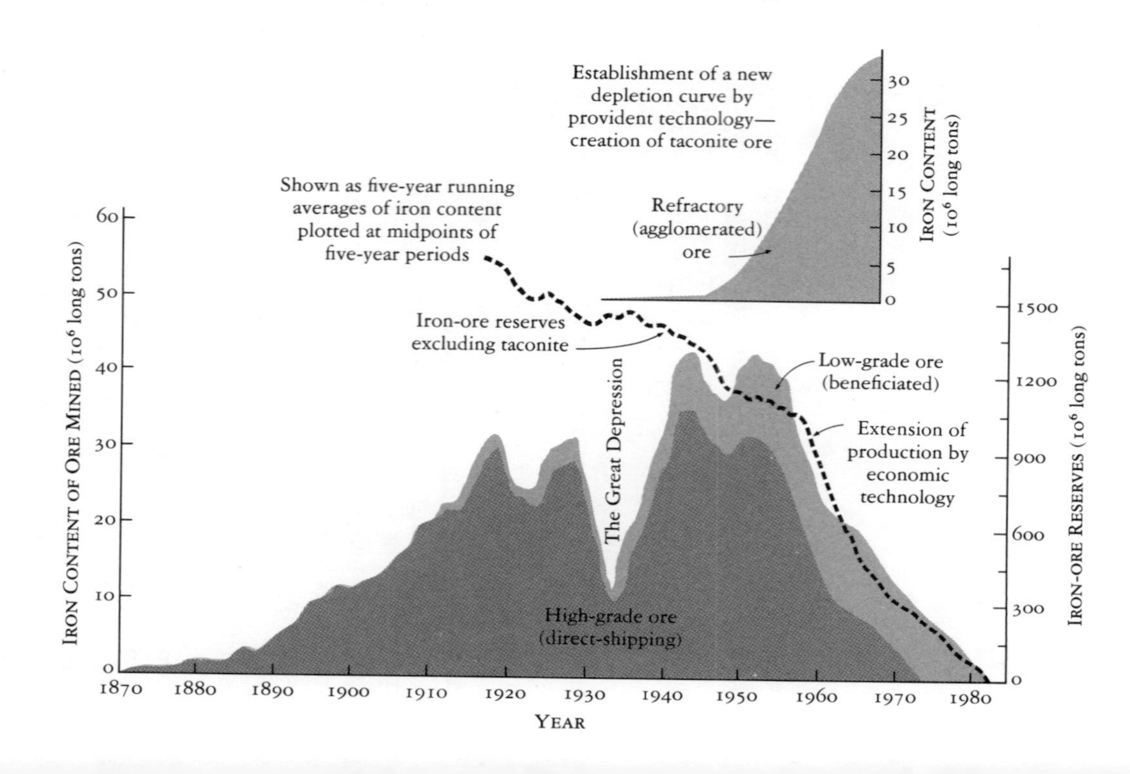

correct, then when we address the question, "Will technology be an integral part of the United States economy?" the key issues must relate to the questions of economic and other resource management policies. Resource management is broadly defined to include the effective use of our material, energy, human resources, taken collectively. All ideologies, all human values, all cultural values, in the final analysis, whether we like it or not, must be found within the sphere of what is technically possible to achieve through the management of these resources.

It is, therefore, appropriate to begin with an examination of the physical and ecological constraints that will delimit our resources in the decade immediately ahead. The most serious concern is that we are running out of time. Time is essential to make the adaptations, changes, and the adjustments. The future need not be dreaded if we have sufficient time to make the necessary adaptations. Thus, let us first examine the physical constraints in which the adaptations in technology and cultural values ultimately must take place.

I shall begin with a few concepts that are central to assessing the time that we have available to make the technological and cultural adaptations.

One of the most important concepts centers in the fact that we do not simply run out of resources in the sense of conventional wisdom. Rather, we go through a production cycle, and these production cycles occur not only in regard to energy but to all our material resources. Figure 1 shows a typical production cycle of iron ore in the Great Lakes region. Iron ore in the Great Lakes region is considered exhausted when the cost of recovery becomes prohibitive. Note that we never developed the tail end of the production cycle because the cost of recovering the ores of reduced quality exceeeds that of recovering taconite—a more energy-intensive resource. Now let's take a quick look at the situation in regard to petroleum. M. King Hubbert, in 1956, applying the concept of the production cycle, predicted that the peak of United States petroleum production would occur in 1973. At the time, this was devastating to the petroleum industry because that meant a decline in the rate of growth. Domestic production did peak out in the early 1970's and even with the Alaskan slope, it will be downhill somewhere after 1980 or before. A friend of mine in Grand Rapids recently made an estimate, a very careful estimate, of what's likely to happen in the next ten years. His assessment shows that even with what we can expect to import into the United States by 1979, we will probably be deficient in petroleum. So we are not talking about the next generation, but we are talking about us now. And if you can imagine the circumstances that might prevail in the Midwest if we have a shortfall in petroleum at the same time we have a shortfall in natural gas, we can see why many of us experience sleepless nights occasionally.

In 1962, Hubbert also gave an estimate of the total production of natural gas. As in the case of the previous production cycles, the area on the curve (figure 2) represents what the geologists estimated as resources in place within the geosphere. The two curves show the range of variation in these estimates—the optimist and the pessimist. It doesn't matter whether you are an optimist or a pessimist. It only makes a few years difference. Figure 3 shows a recent analysis of the situation as presented by ERDA. Note that natural gas we hope to develop in Alaska will help out for a little while, but it's inevitably down after about 1985. Figure 4 shows the expected production cycle for global production of petroleum. If

FIGURE 2

1962 ESTIMATES OF ULTIMATE AMOUNT OF NATURAL GAS TO BE PRODUCED IN CONTERMINOUS UNITED STATES, AND ESTIMATES OF DATE OF PEAK PRODUCTION RATE

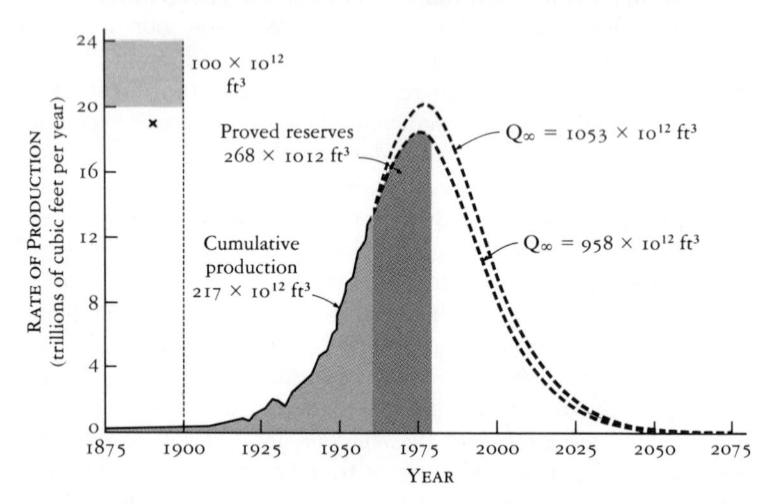

SOURCE: Redrawn after Hubbert (1962).

FIGURE 3

PROJECTED DOMESTIC NATURAL GAS PRODUCTION

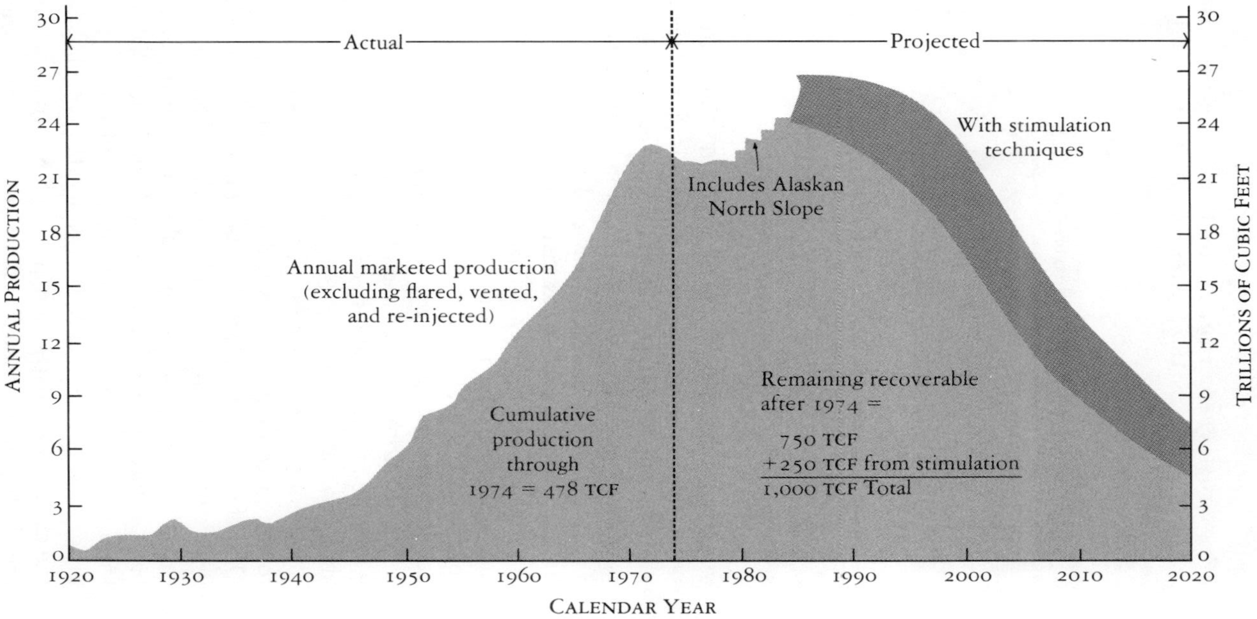

FIGURE 4

1956 PREDICTION OF THE DATE OF PEAK IN THE RATE OF UNITED STATES CRUDE-OIL PRODUCTION

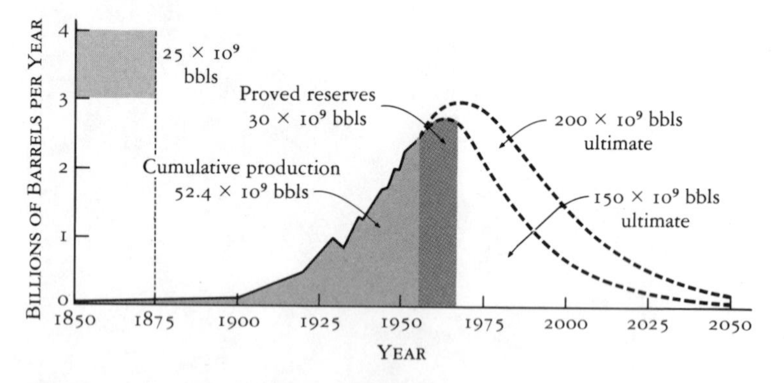

SOURCE: Redrawn after Hubbert (1956).

you're an optimist, the world production in petroleum will peak out in the year 2000; if you're a pessimist, it will peak in 1985. Also note that the human species will have used up eighty percent of the petroleum on the face of the earth in 58 to 64 years.

It must be emphasized that it takes energy to get energy. This first half of production was easy to recover, the second half will be very difficult to recover. Consequently, the *net* energy that will be available from the remaining petroleum will be significantly less than that obtainable from the first half. We will probably never develop the tail end of the production cycle.

The economic impact starts when that curve starts bending over, when the rate of growth starts to decline. This is expected to occur about 1979.

One of the important aspects of technological adaptations relates to the price of petroleum and how rapidly it changes with time. One of the remarkable things about the industrial revolution is that it gave us a capability to centralize our production processes and to capitalize on economies of scale. In the extraction industry you have a third dimension to that problem called "entropy." Derived from the Greek word for change, it reflects the fact that it takes progressively more energy to extract the remaining resources from the earth. In the extraction industry, the economies-of-scale curves are all displaced upward as illustrated to form a three-dimensional

FIGURE 5

TECHNICAL COST OF RECOVERY AS A FUNCTION OF PRODUCTION RATE AND CHANGING ENTROPY OF RESOURCE

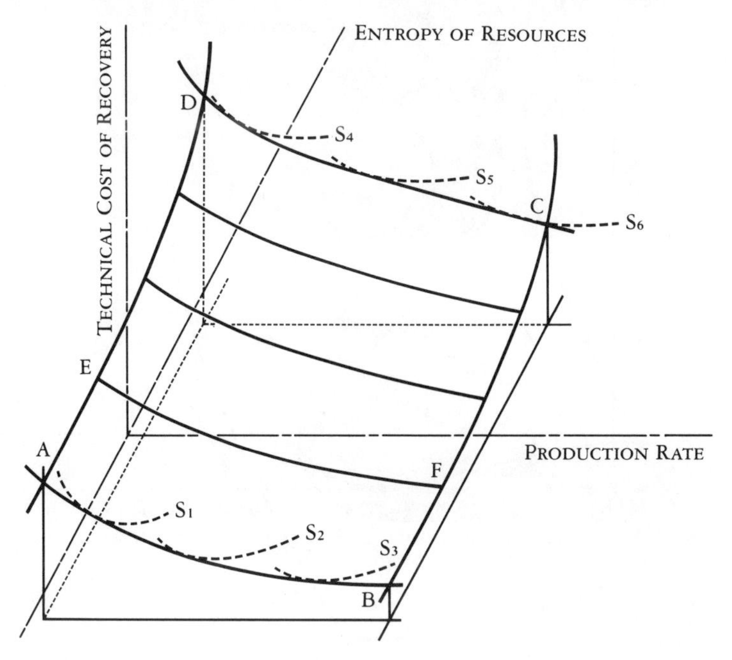

surface as indicated in figure 5. Now let's see what we can expect in the way of change in the real costs of recovery, as one progresses through a production cycle. The results are illustrated in figure 6. The exploitation of petroleum begins at low volumes. The volume of production increases as centralization of industries occurs and the cost decreases as illustrated. Petroleum-based fuels we buy today cost less in real terms than what they cost in the 1930's, for example.

We have been walking down the hill, so to speak. After production peaks out, you start climbing the cost hill. Not only are recovery costs increasing in real terms but the rate of production must decrease. It costs fifteen times as much to recover a barrel of oil out of the hostile environment of the North Sea as it does in the Middle East, measured in real cost. In addition, the investment risks initially are enormous, and huge capital investments are required for recovery. The returns on these capital investments must

FIGURE 6

ADMINISTERED PRICE AS A FUNCTION OF PRODUCTION RATE AND CHANGING ENTROPY FOR A FINITE RESOURCE

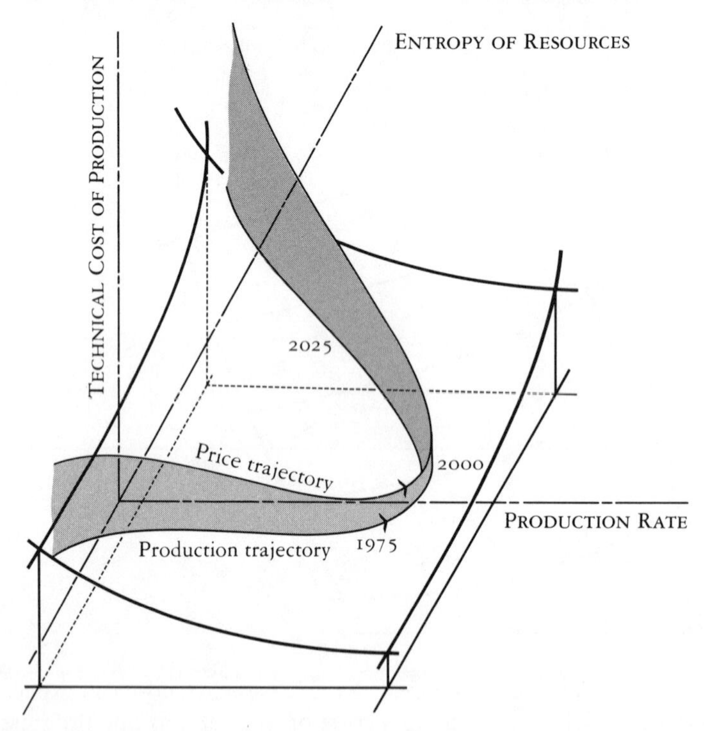

be spread over a progressively lower volume as you move down the production scale.

Note also that in order to recover the resources at all, very large capital wealth and resources must accumulate under the management of a few people. The dynamics I have just outlined explode the myth that small oil producers can get the job done. Another myth that pervades our society is that we have lots of coal and all we have to do is shift to coal. Figure 7 shows the production cycle on coal in the United States. The early stages of the Industrial Revolution in this country were supported by coal. We shifted to fluid fuels for very good reasons. All we had to do was poke a hole in the ground and suck the fuel out of the ground like a soda straw. Further, we pump it into homes all over our nation and burn it without ash and under the control of a device called a thermostat.

FIGURE 7

COMPLETE CYCLE OF UNITED STATES COAL PRODUCTION FOR TWO VALUES OF Q∞

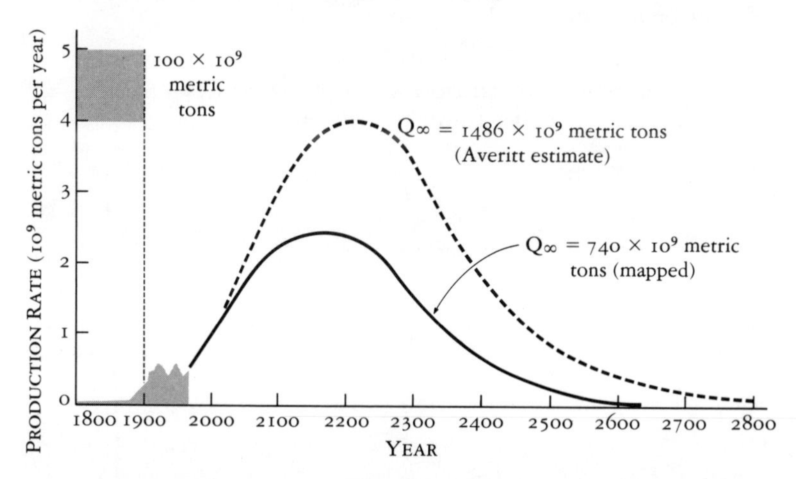

Now we frequently hear people say, "Well, we have two hundred or three hundred years of coal left." But, as M. King Hubbert points out, the area under the curves in figure 7 includes coal that is 14 inches thick and 6,000 feet under the ground. If you consider the coal that is 28 inches thick and only 1,000 feet under the ground, you must reduce the area of these curves by a factor of 4.

FIGURE 8

EPOCH OF FOSSIL-FUEL EXPLOITATION IN PERSPECTIVE OF HUMAN HISTORY

(from 5000 years in the past to 5000 years in the future)

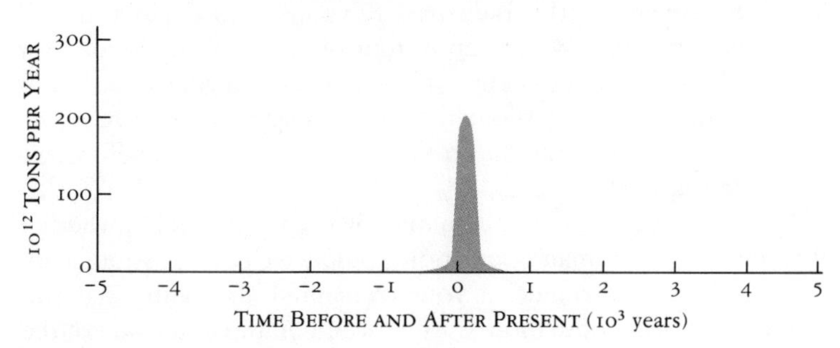

SOURCE: Modified from Hubbert (1962).

There is no question that the fossil fuel era will be a mere "blip" in the history of mankind, as shown in figure 8. We are virtually at the top of the blip. Will there be technological adaptation? You'd better believe it.

FIGURE 9

ENERGY CONSUMPTION AND POPULATION TRENDS FROM 1 A.D. TO THE PRESENT

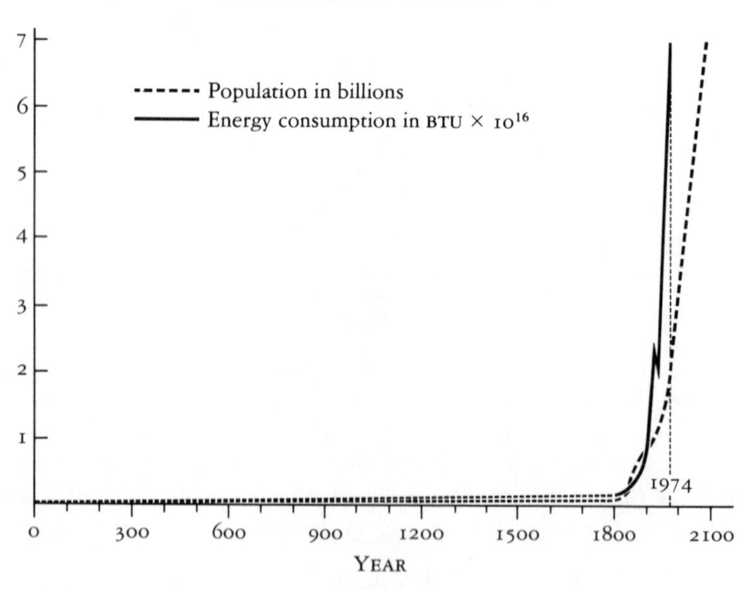

Figure 9 shows an expanded scene of the blip with the simultaneous growth in human population. Prior to the Industrial Revolution, human population was hardly visible as a line in this diagram. With the advent of the Industrial Revolution to exploit fuels it grew geometrically. We are up to four billion now and headed for eight billion in another twenty or thirty years if all goes well. Biologists tell me it's not as though the human species didn't have the biological capacity for an outbreak in human population before; we just didn't have the energy.

You can struggle in your own mind with the question of whether this outbreak in human population would have been possible without fossil fuels. Struggle in your own mind also with what will happen to world population as we descend on the other side of the production cycle.

FIGURE 10

NET ENERGY CURVES

(Corresponding to different energy ratios E_R, for a building schedule with a doubling time of five years)

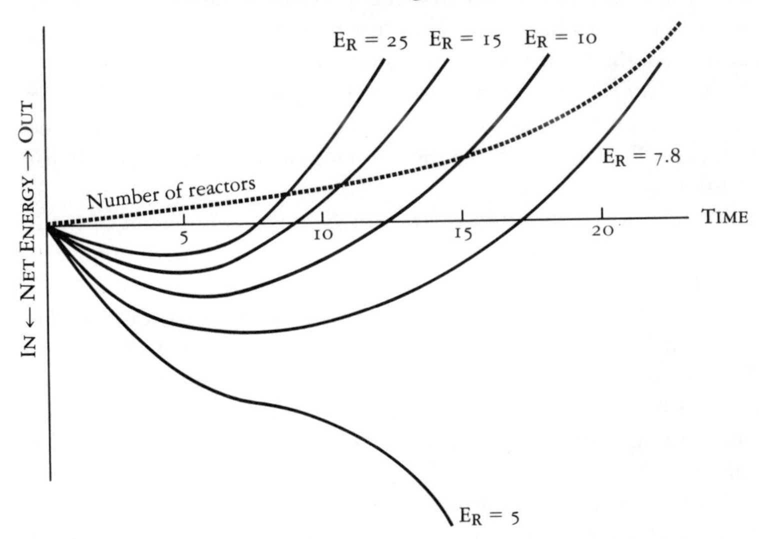

Some say that we can and must make up for the decline of fossil fuels with nuclear power. There are all kinds of problems with nuclear power socially, technically, and environmentally. I'm not going to go into those. But I do want to point out a sobering fact. There is no way we can develop this and other resources fast enough to "outrun the problem" even if we made an all-out effort to do so. The problem relates to the important concept of energy ratio. The energy you invest in the development of a nuclear facility and in operating it over its lifetime compared to the energy derived from the facility during this time is called the net energy ratio. In the case of nuclear power plants the lifetime is about thirty years. The net energy ratios have been calculated to be somewhere between less than one for low-grade uranium ores and up to as high as 19 for some high-grade ores depending also on the technology used, as indicated in figure 10. An energy ratio of 10 means that you get ten times as much energy out of that facility over its lifetime compared to what you invest in it.

Now consider the dynamics of developing nuclear power systems. If one builds nuclear plants at the rate of one per year, you invest for a while and then ultimately begin to realize a return on your investment. The program of nuclear development in this country at one time called for a growth rate of 15 percent per year. That's a doubling time of five years. Figure 10 shows the energy return on energy investment as a function of time. If the net energy ratio is 10, the return is negative for about twelve to thirteen years. A net energy ratio of 5 at that development rate would never provide a return. There are intrinsic time constraints that cannot be removed. This is why I put such an emphasis on the time scale for change.

Insofar as solar energy is concerned, there is lots of it but it is very difficult to transform into work. It is the work component of energy that runs the wheels of industry, and it is the work component that gives us mobility. Suppose you had to build several solar collectors with the energy you derived from your solar collectors. That would be very difficult because you could convert at best only a very small portion of the energy you collect to a form suitable for smelting and processing the materials required to build the solar collectors.

There are, however, forms of solar energy that are very well adapted to doing work. Wind and hydro are such examples. But I see no way that this society can possibly maintain the high-energy-intensive economy we now have on solar energy. Adjustments in our way of life, our technologies, and cultural values are required for a solar-based economy.

Now, what are some of these adjustments? Undoubtedly the first thing we will try to do as the real cost of energy increases and it becomes less available, is make adjustments to more efficient technologies, thermodynamically speaking; they are frequently called "technological fixes." We can and will put more insulation in our homes, opening windows a little more in the summer, turning down the thermostat in winter. Detroit will build a more efficient automobile for you, you'll get 45 or 50 miles per gallon instead of 16 in a few years. We will transfer some of the freight from the highways to the railroads. Such technological adjustments have been estimated to have the potential of saving thirty to forty percent of our present energy budget. Such adjustments will not affect our life style very much, but the "slack" will run out after a few years.

The next large class of adjustments relate to our mobility. Fifty-four percent of our petroleum is used for transportation. As petroleum becomes less available, and the price rises significantly, our mobility will go down. How will this affect the landscape? We will undoubtedly find that what we need to re-invent are medium-sized communities built around a small electrical generating facility that heats our homes and commercial buildings with the residual heat. It is called district heating. These communities would also serve as transit terminal connections to other communities, and they will have a diversified commerce and some decentralized, light industry. Such communities are much more energy-efficient, since most of the elements of everyday living are near at hand.

A third class of adjustment is to be found in the area of product durability. Just as small is beautiful, age will be beautiful. We have the technology to significantly increase the durability of most of our products. We can build a refrigerator that will outlive its human owner. Detroit is beginning to realize that it should develop cars that will last perhaps twice as long as present cars. If we increase product durability, as we must, there will be a tremendous opportunity for not only reducing energy requirements, but also the impact on our environment. The side effect of the economic impact is potentially enormous.

The impact the increased product durability will have on the employment in Detroit, however, must also be considered. We have depended explicitly for the last thirty or more years on the ability to expand production as is necessary to provide jobs. That day is gone. We will have to find other mechanisms for providing employment.

We moved thirty million people from rural America to urban America in the thirty years following the Second World War to man the wheels of industry, which we kept speeding up to manufacture more and more. We probably will once again rediscover small-scale diversified agriculture employing low capital, a decentralized, more labor-intensive agriculture.

Those are the technological adjustments I see forthcoming. I do not perceive the future as dismal. As a matter of fact, I look forward to it. The only thing that worries me is the transition problem.

Changing the Scale of Technology—Paths to the Future

Louis B. Lundborg

I want to divide my presentation into two parts. First, I want to address briefly the topic that is assigned here today: the impact of the changing scale of technology, the implications for labor, management, and economic growth; second, I want to focus for a few moments on something that I think is not adequately stressed anywhere in the agenda of the several parts of this conference. In both parts of my remarks, I want to draw a distinction between the Third World developing countries on the one hand, and the developed world, the Western world largely, on the other. While the purposes and objectives of intermediate technology are exactly the same in either one, the methods and approaches to it are almost totally different. It is relatively easy to apply the principles of intermediate technology or appropriate technology in a geographical area that is not yet saturated with inappropriate technology and is still starting more or less from scratch. At ground zero it is infinitely easier to apply the things that are being talked about here than it is in the already developed countries, the countries that are already overloaded with inappropriate technology. But by far the biggest problem, in one respect, is in the developed countries. And until this session, I think most of the focus has been on the developing countries. But in terms of what Professor Koenig has said about the consumption of energy and of all kinds of resources, the problem is infinitely greater in the developed countries because that is where the consumption is just geometrically greater than it is now or is likely to be in the developing countries within a foreseeable time.

Now first addressing the three parts of this morning's agenda, the implications for labor, management, and economic growth in changing the scale of technology. First, the matter of labor: it can be tremendously disruptive unless we plan and implement a transition. I could dismiss the whole thing by saying the same thing about all three components. All these elements of our economy could be tremendously disrupted by the application of the changing scale of technology unless we prepare for it and we handle smoothly a transition resulting in just that, a transition instead of an abrupt change. It can be felt most keenly in the area of labor, and I don't mean organized labor; I mean personal, individual, human labor in all its forms, organized or unorganized; it can be shatteringly disruptive or it can be clear at the other extreme, highly rewarding—again depending on whether and how a transition is made. It can be rewarding in the sense that the kinds of things that people can do in a different mode of technology can be infinitely more rewarding than some of the things that are done now with high technology, at the lower end of the labor scale in a high-technology economy. The things that can be done in the area of repair, rehabilitation, recycling, the restoration of soils, the reforesting, that entire area of things calls for the degrees of skill, types of skill, types of involvement that can be infinitely more rewarding if they are implemented and done. But the whole area of labor can be the biggest disruption if it is not properly handled.

The implications to management are that the transition can be disruptive. I would say for my fellow managers that they aren't good managers if they aren't able to accommodate better to change than labor is likely to. Management has more of an opportunity to adjust; but again, only if a number of things happen that are transitional. One of the pressures that brings about the problem that we are all agonizing about now, is something that has been imposed upon management—industrial management, corporate management—by a force we've hardly talked about in this conference. That is the investment fraternity. Managements are judged by the investment fraternity on the basis of something called "performance." And performance is equated with growth. The investment world itself has changed within recent times. We talk about 25 million owners of investment securities in this country, but the great bulk of all the ownership now is in large institutional hands and largely in pension funds. They are held for the benefit of workers for future pension benefits, but they are managed by a group of investment managers in these institutional funds which are largely pension funds, and some of them are mutual funds. The principal criterion those investment funds have been using for measuring whether they do or don't buy, or do or don't hold stocks has been what they call "performance." And that in turn, as I say, largely equates to continuing compounded growth. As one component of transition we are not likely to take the pressure off the industrial part of our economy until we can somehow persuade ourselves as to our own part of the job. None of the problem we are talking about is going to be accomplished by "them." It is going to be accomplished by "us," somehow or other. Part of our problem is to find, I think, where we can get a handle on the things we have to do; and I think one of the things we somehow have to accommodate ourselves to is an investment world that doesn't put the pressure on continuing compounded growth.

The third element on our agenda says "economic growth." Again I will quickly capsule what I think will be the impact and then if there is time, we can take it apart in pieces. We hear about zero growth, we hear about limited growth, we hear about restrained growth, but we don't define the terms very well. I am convinced that one of the impacts, one of the implications, has to be reduced industrial growth in those areas of industrial operations that are based on the consumption or the use of non-renewable finite natural resources. There is no alternative but restraint, probably a downward curve in that segment of the economy. But when we talk about economic growth I think we should remind ourselves that economic growth, economic affluence, if you want to use that term, is not limited entirely to industrial growth. It is not limited entirely to the production of things that use up natural resources. We are no longer just an industrial economy; we are now, for the first time, since the very last few years, a different kind of economy. We have gone historically from being a predominantly agricultural economy to being a predominantly industrial, manufacturing economy. Now we are neither of those, rather we are now a service economy. There are a great many things that should be and properly are measured as part of the economy and part of economic growth that do not depend on using scarce natural resources; and if we are ever going to accomplish the transition, it is going to have to mean a shifting away from the resource-using industrial activities over into the non-resource-using service activities which can be highly rewarding to human life and human enjoyment. If you define economic growth in those terms, I would say we could have continuing economic growth. If you define economic growth only in industrial, resource-using terms, the implication of changing technology is a downward trend in that kind of economic growth.

Now, having addressed myself, at least in a capsule way, to those elements of our program, let me turn to something else. In this kind of world, people are inclined to wring their hands in despair over these problems we have been talking about; to feel unable to cope with them; or simply to dismiss the problems out of hand as just not being serious enough to justify concern. One of the things I think Dr. Schumacher has been doing for all of us in the whole world is that he has been putting some of these major economic social dilemmas into perspective, into a framework where that can be discussed and can be attacked rationally and pragmatically. But it seems to me that the element that has been missing in most of the discussions of intermediate or alternative technology and in most of the other discussions about necessary restraint in the rate of industrial growth, has been any suggestion of how the transition is going to be accomplished. No one has talked concretely or procedurally about how we get from here to there, even among those who favor the shift or who recognize its inevitability. I count myself in both groups: I favor personally as a matter of my own human, sentimental likes and dislikes, some shifting away from what we have been doing, even if we did not have to do it, but I think we have no choice. But I am not unhappy about the inevitability of it. Among those in either of those camps, there is a pattern of saying, "Well, this is the way it ought to be," without anything like a blueprint or a roadmap specifying the steps that have to be taken between here and there.

One thing I have been glad to note is a focus on how very recent and new some of these things we agonize over are. Some of my business friends get almost apoplectic about what they see as threats to what they call the American way of life, free enterprise, things of that kind, not realizing that many of the things that seem most sacred to them are less than a generation old. Some of them are post–World War II, and yet they become very defensive about these things. This whole production, consumption, distribution syndrome we are in the midst of is a very new thing. We cannot put an exact year on it. Some of it started back in the twenties and thirties; largely, however, it became a total thing after World War II, this matter of our being able to produce more than was truly essential for ordinary needs. So it became necessary, in order to keep the economic wheels turning, to push things out in a distribution kind of way. This is a brand new thing and yet now it has been so built into the system that, of course, people always want to defend the thing that is familiar to them and that they depend on for their own livelihoods. I don't blame people at all, and I think you have to get that into perspective. We do have to remember also that not only are they dependent on it but our whole system of employment is built around it. So if we are going to move away from it we have to have something ready to take its place.

Now I realize that there are boobytraps and there are land mines any time a specific plan or a specific model is set up. The fate of the Club of Rome's *Limits to Growth* is a good example: a model became the target for everyone to shoot at. And yet the very absence of any kind of approach often has become equally a target, or probably more accurately, it becomes an excuse, it becomes an alibi for no action. Because it is so hard to chart and because any path seems so strewn with rocks, roadblocks, and with pitfalls of one kind or another, people are inclined to say, "Let's just rock along, let's see what happens. It may not be as bad as it seems and anyway we have always found a solution before." Well, that just isn't good enough this time. We already have too many signals that the old pattern of just letting things work themselves out just is not a safe one to trust. So at any conference like this, I think we need to do a little intermediate technology of our own. We cannot hope here to devise a complete model with timetables, flow charts, critical-path methods for every step of the way. But we should at least begin to identify what some of the steps are; and for starters, let me just suggest a few kinds of things. Last night Dr. Schumacher said we should not leave here unless each one of us could say to himself, "What can I do?" Well, I feel very strongly about that and I think the "What can I do?" should be pointed toward doing something specific about the problem, and particularly about the transition. So let us talk about some examples of the kinds of things that everybody can begin to address himself to.

First, let us review all of the legislation, legislative enactments, the regulations, whether they are legislative or regulatory, that encourage wasteful growth of any kind. Start with things that encourage population growth—and our law books are just full of them—the premiums, the bonuses on population and other kinds of premiums on using up things.

Let us address ourselves to revision of investment standards to de-emphasize excessive growth as a key measure of performance. I daresay there is hardly a person who is not somehow tied into some kind of an investment fund. The faculty are part of TIAA or some other teachers' retirement program. Others are sons and daughters of people who are presently or potentially in retirement, and so on and so on. All of those are funds being invested and managed by professional managers in institutional funds of one kind or another. We can begin to feed back to the people managing our funds that we want investment measured by something other than growth as an index of performance.

We should identify critically needed public projects that would restore resources and at the same time would be labor intensive. This is critically necessary to absorb the large numbers of workers who are going to be displaced by any shrinking or leveling of industrial production and yet it can be done. But, it will only be done if it is addressed as a specific target objective.

Also, we must identify service occupations that would be socially and economically useful but have been avoided because they lack prestige. Along with that, we have to develop training and other programs that would create pride in those occupations. Let me give you an example. There is a crying need in this country to have aging people cared for in dignity in their own homes instead of shunting them off to demeaning nursing homes. And yet, it cannot easily be done now because that sounds like domestic work and "domestic work is demeaning." The devil it is demeaning! It can be the most gloriously dignified, rewarding work in the world; if it were not, you would have no nurses; you would have had no Peace Corps; you would not have "Amigos de las Americas." The caring for human needs can never be undignified. It can never lack prestige if we get rid of the silly notion that there is something demeaning about helping other people. But a lot of things have to be done so that it does not become demeaning. There has been a history of exploiting domestic help. It should not be. I think we need to set standards there so that whether people work for individuals in a home or do this as an organized thing to help aging people they will be decently treated in their employment.

These are only a few of the many kinds of people who need help in their own homes; it is a tremendous field. I mentioned the Amigos; you who are familiar with them know that they are one of the several paramedic programs, most of which have been carried on abroad in the developing countries. The present system of the delivery of medical care right here in our own country is woefully inadequate and is getting worse. There is tremendous need for many levels of paramedic systems.

As I mention these things, let me stress that this list is not intended to be definitive; it can be expanded indefinitely. The things I mentioned are not even in any order of priority but they are thought-starters to indicate the directions that I think we should not only be thinking about, but should be starting to act on; to act right now. And with that may I close by paraphrasing the Book of Proverbs: "With all thy getting, get going."

Automation and the Scale of Technology

E. F. Schumacher

A recent news release from America's most respected technological market research organization says that the United States baking industry is revolutionizing its processes, and sees automation as a most likely solution for its progress and prosperity. The supermarket will simultaneously become the industry's largest retail outlet for eighty percent of all bread sales. It goes on to state that the overall move will be towards fewer baking plants with larger processing facilities, distribution ranging up to five hundred miles from base, and a marked decrease in bakery personnel. The study documents a required turn towards highly automated equipment producing over fourteen thousand pounds of bread per hour per unit; profit and growth will depend on the creation and utilization of higher speed process flows. Present packaging machinery running at the rate of 60 loaves fails to compete with lines capable of producing bread at the rate of 240 loaves per minute.

So let us not assume that we are in a stable situation. We discover that technology is running away from real human concern, and a little slogan like "small is beautiful" may help to persuade people that this headlong rush into giantism is the road into perdition.

How can we adapt ourselves to the things that are coming towards us? In the title of this discussion there is the word "labor." Well, at present in the Western world we have fifteen million unemployed, and the general attitude is still that the next boom, the next expansion, if it should come and if it did absorb them, would immediately produce the second really severe fuel crisis, and throw us back again. It is really not very intelligent to discuss the desirability of growth, particularly if nobody tells me what is supposed to be growing. I mean, when my children grow it's a good thing. If I should suddenly start to grow, it would be a disaster. But people just handle these words as if they were realities. It's not very intelligent to discuss abstractions when it is important to determine how we are going to prepare ourselves for the future.

I don't think the future will be masterd unless we become conscious of what I call "people's power." The people can do a great deal. I entirely agree with Mr. Lundborg that we have ensnared ourselves in so much petty legislation that people's power is normally inhibited. So that has to be battled with. That petty legislation has to be removed. There has to be a dash for freedom in this respect. The people can do a great deal for themselves (and, please, I don't want to be accused of heaping additional burdens on the women; I bake the bread at home, not my wife); I mean something much more general. The problem of medi-care was mentioned a few minutes ago. In my country, we've had an excellent National Health Service since 1947. But a strange thing has happened. People now imagine that it's the task of the medical profession to keep them healthy; that they can adopt the worst possible diet and denounce everything that is sane as crankiness, because, after all, it's not *their* task, but the doctor's, to keep them healthy. Not only in the United Kingdom but also in other so-called advanced countries, there is an immense increase in the proportion of national income that has to go into medical services. If you project this trend only a few years, you find that all the national income will be needed for them and then there will be no more need for any health services. This also has to change, and it is something that we can do. We can understand that it is our task to keep ourselves in good health, and this is one part of the revision of life styles which is already coming. It is one of the corrections where we see that people have not entirely lost the power of helping themselves. To be afraid of these changes, primarily because there will be shifts in job opportunities, well, this is a kind of "paralysis by analysis." If we wouldn't analyze quite so much we might act more effectively; because it's not very difficult to understand what we ought to be doing.

Working with and through Institutions

Sim Van der Ryn

Tom Bender helped us see that "appropriate technology" is really a clumsy way of saying "right livelihood" or "good living." Dr. Koenig brought us face to face with the fickle finger of fossil fuels. I like that symbol a lot. Think of it as a hypodermic needle that shoots us full of the fossil fuel addiction that we're trying to move away from. Mr. Lundborg, I think, touched on the heart of many issues. From Dr. Schumacher's last comments, the thing I find most important is that the real opportunity for change is to occur in two areas. One, in moving away from that fossil fuel fix towards building sustainable society by remodeling what we have now since we can't just throw away our existing stock of buildings. The true growth industry is moving from a mechanical to a biological age. I guess the two things I might consider, because they're things that I do know something about, have to do with skills and how we do some of this. The other one, which has been alluded to a few times, is called "institutional impediments to change," which I'll call "hassles." Let me read you a little hassle. The *Sacramento Bee,* from the summer of 1976:

> The Dunsmuir Leather Works was crippled by a state law forbidding women to work in their homes on the firm's products and has closed their stores. It once had a payroll second in the city only to the Southern Pacific Railroad.... The beginning of the end was January 1st when a new State law went into effect that put teeth into the State's old anti-sweat shop regulations. The new rules made illegal the unsupervised piece work the local women have been doing in their home for

> Dunsmuir Leather Works. Some of the women were making as much as $100 per week for what they considered part-time work.

Now this is absolute insanity. All those women are back on welfare now. What this fellow did was simply have women do work in their homes on buckskin garments, very fancy buckskin garments. They worked on their own time. As some of you know, Dunsmuir is a fairly remote, rural area, with not much employment. As a matter of fact, the only thing that's there is a railhead. So, here we have a law which is designed to protect newcomers from Hong Kong and Mexico from being exploited in sweatshop conditions in Los Angeles and San Francisco, and it wipes out employment for rural people in northern California. So, this is the uninformed uniformity of regulation and law.

Let's consider skills. Let's consider pension funds. I recently met with a group of Vietnam veterans. There's a surplus of money for those veterans to go buy tract houses and get VA loans, but there's no money for those veterans to go out and start businesses. They were up in Sacramento to see the governor about how some of those funds which are now very poorly invested at a very low rate of return could be used so they could get involved in appropriate technology. Maybe we should have a law that a certain amount of pension funds are invested in socially desirable purposes to help us make the transition to a stable economy.

Now a few things about the educational process. My experience is, we don't have people with the real kinds of skills that are needed. I think we can get them rather rapidly. I think it means making more sense out of our educational system, particularly at the university level. Let me give you a few examples. My friend Wendell Berry, the poet and farmer, also happens to teach at the University of Kentucky. When he was here recently, he came over to the campus and said, "Why don't we cut the salaries of professors of agriculture in half and give them a piece of land to farm?" When I taught architecture at the University of California, we thought we'd build a house. We put in for a research project, and with some students we'd work together and build an energy-efficient house. Well, it never got out of the research office. They said, "You can't have students build houses. The unions will be all over us." So what I did was raise the money myself, and we went out to the engineering field station, which was rather vacant, and we figured if we weren't visible maybe we would be left alone. We started to build right there, and right next to where we were building were the university shops, where the carpenters and the steamfitters work. They became the teachers and we got to use their shops.

I have spent a lot of time in the last five years trying to make some sense out of our waste management practices. You look at the curriculum of sanitary engineering in places like the University of California and there is not one course that deals with on-site systems. Thirty percent of the country is serviced by on-site systems, particularly septic tanks and leaching fields, and these are complex biological systems that are very simple from an engineering point of view, so they are not of very much interest to sanitary engineers. Here at Davis, a big agriculture center, do you know what happens to the manure from these barns? It isn't composted; it goes to the dump! If the sixties was the era of civil rights where we tried to gain rights and equality for people who traditionally have been denied them, the seventies and eighties are going to be a time for fighting for our individual rights to build a good way to live. Now what I've discovered—it's not a conspiracy, and it's not even a bureaucracy—is a simple principle that large units like to deal with other large units. So in order to get smallness, we're going to just have to struggle and struggle to make some sense out of all this.

The issue of safety has been touched on. You know we have been bankrupting ourselves by trying to indemnify every possible condition under which we live. Farallones Institute started trying to work directly with people in institutions to make some sense out of their environment. And some of our first work was in school playgrounds. Our main problem was insurance. It's fine to let children be on a prison-like school yard covered with asphalt and maybe a jungle gym or two, but put things in there that involve body motion and kinetic experience and so on, and right away you couldn't get insurance. Well that's still a great problem. Some districts that have really living playgrounds have had their insurance cancelled.

In the area of building codes which I have had experience with, change is slow to come. You still cannot use recycled lumber in building a house, according to the Uniform Building Code. You can't use lumber that you've milled yourself for building a house. The lumber has to be graded by a person who comes and grades your lumber. Well, I suggest to people: go make up a rubber stamp, a grading stamp, and stamp your lumber. We have to find practical solutions to practical problems!

One truly excessive uniform set of laws is the famous occupational and safety hazard regulations (OSHA). Now there are many problems with industrial safety, and our health is jeopardized in many instances. But what the net effect of all this regulation is, is to hurt small business and large business too. You take a self-employed person having a small automotive repair garage. Now for someone to have a garage, they have to hire a bookkeeper just to fill out the forms because now, you know, we have new consumer laws to make sure that you don't get gypped by your local garage. If we had healthy communities you wouldn't be gypped by your local garage. So, we have government. Let us build healthy communities. If we don't there won't be anything left. We'll just have more and more laws to try and indemnify the fact that we don't have healthy communities. That is a very basic problem.

Let's talk for a moment about OSHA. The Libertarian Party last year did a very nice poster of what a cowboy would look like if he met all the OSHA regulations. You know, a Marlboro man with rollover bars and a cow catcher and goggles. The horse on rollers, so he wouldn't fall. I heard of a fellow in St. Louis that had a small manufacturing business, with rather old but very durable machinery, and the way he got around the OSHA inspection is he had all these wonderful signs that said, "Dangerous machinery, do not use, do not touch." And when the OSHA inspector came around, of course, he just closed the shop and hung these signs on all this equipment. There are ways. We have to develop an intelligence system, you see. A little intelligence, a little communication goes a long way.

In setting up our rural centers in Farallones Institute, where we are trying to develop and live and research an ecologically sound, right way to live, we ran into all kinds of problems. We went to the county initially with a plan for the use of a piece of land but it wasn't five-acre ranchettes, it did not give everybody his own compactor and kitchen, and so on, because we had grouped some of our community facilities. They say, "Ah ha, you're an organized camp." We said, "No, we're not organized, we're not the boy scouts, sorry." They said, "Well, you're a school." We said, "We are a school too. But we're more than a school, but besides we don't want to be zoned as an institution which puts all other kinds of requirements on us." Well, it was a very interesting process and we didn't fit into any category and we still don't. They finally decided we were a kind of school you see, and then there was a set of health and safety laws that apply to schools.

Now, for example, as part of our teaching program we grow our own food. We have our own animals. We can our own food. Well, on the first visit from the health inspector, he looked in the refrigerator and saw some bottles of milk. And he said, "Where did this milk come from?" I replied, "From our cow." Well we had a certified letter the next day. "You are in violations of the Health and Safety Code for drinking milk from an unauthorized source." Our kitchen was technically a restaurant and we were required to have a changing room for our waitresses.

Well, it's all amusing now; it wasn't so amusing then, because they can turn you out of your home for doing things that make sense. It's still happening and it needs to change. And maybe today we can get a little further towards making it change.

Discussion

AUDIENCE PARTICIPANT: Basically it seems like the most difficult aspect of this transition stage is trying to convert a society that can't distinguish between wants and needs, and between rights and privileges in the case of automobiles. What I'm wondering is how can this transition take place without it being a threat to personal freedom when you have a situation like in the state of California where people won't voluntarily cut back on their oil? We're probably going to have to reach a point where there is going to have to be compulsory allocation of resources, and I'm just wondering how we are going to redefine what personal freedom means when people don't understand that driving a car is actually a privilege instead of a right.

LOUIS LUNDBORG: The problem, it seems to me, is that we have false impressions of what our freedoms are. In short, the box we live in is a little tighter than we thought. So this means that the only way we can make that transition, it seems to me, is with a massive educational effort so that we all understand what the realities of the situation are that we are trying to deal with. That's the only thing I can really say.

AUDIENCE PARTICIPANT: Why is this such a homogeneous group? Is not the concept of small is beautiful for poor people? If so, why aren't they fully represented? A question is quickly raised in my mind. Is the concept that appropriate technology is for poor people another myth for the middle class and thus another victory for the ruling class and their institutions? I think it depends upon interpretation. If we think that small is beautiful or appropriate technology is a rationalization of stifling such obviously bullshit programs such as the SST or space program, or to save energy, or to get a handle on run-away, large-scale technology, then the concept of small is beautiful is a myth. It's a distortion. You are missing the point. Our institutions need to be changed. The institutions are on an inappropriate scale. They are controlled by too few people. The people, especially the poor people, have no control over the institutions and quality of life the institutions dictate. The system must be changed. Small is beautiful must be seen as a method for people. Local people to control the institutions, to run the system, and thus their lives. Otherwise, small is beautiful is a cop-out, a security trip for the middle class and another ploy of the institutions and their controllers. The crises that relate to appropriate technology such as resources, growth, and energy, are important. But the beauty of appropriate technology is that these crises will cease as a result of the application of small is beautiful; not attempting to resolve the crisis with the rationalization that small is beautiful. Is the strategy of education enough? Education is important, yes, but is it enough? Do we need other strategies? More effective strategies to change the institutions? I know so. It will take action: political, economic, and social action by all kinds of people. People must unselfishly cooperate and not selfishly compete. If you have any comments I would appreciate them, but otherwise it will stand by itself.

VII

ENERGY AND INTERMEDIATE TECHNOLOGY: QUESTIONS OF SCALE, NEW TECHNOLOGIES, GOVERNMENT REGULATION, AND DECENTRALIZATION

Paul Craig

Steven Slaby

Emilio Varanini

Jonathan Hammond

Kenneth E. F. Watt

Dan Whitney

INTRODUCTION: *Richard C. Dorf*

MODERATOR: *Paul Craig*

Introduction

Richard C. Dorf

One point of agreement in 1977 is that the availability and price of energy is one of the important topics of the decade. Beyond that simple agreement, a widely diverging series of analyses and proposed solutions exist. Among these issues is the possibly more appropriate use of energy and the potential for new intermediate technologies for the production of energy and power and the translation of this power in more efficient ways.

The energy crisis of the 1970's began with the Arab oil embargo of 1973 and continued through Project Independence in 1975 and President Carter's energy plan of April 1977. The energy issue brings the citizen to a crisis in values as well as a crisis in energy supply and cost. Who should have energy, what should it cost and what is a judicious use of the world's limited fossil-fuel resources?

Intermediate or appropriate technology becomes defined as a technology that is less wasteful of energy. As an example, the most appropriate transportation technology in a city like Davis with its flat, well-laid-out streets and bike-paths is the bicycle. In other cities the most appropriate means of transportation may be the rapid transit train or the bus.

With declining resources, unneeded articles of consumption may need to be valued less. Increased recognition of the effect of government regulation on energy will be needed. For example, new effective building codes are needed that emphasize energy conservation. Are the products of high technology, such as the nuclear power-generating plant, an effective investment? Or should the nation pursue the development of new intermediate energy technologies that lend themselves to dispersal of location and use? For example, should the federal government shift its research funds from nuclear fusion and fission to solar, wind, and fuel-cell technologies that can be used by individual homes and neighborhoods? The effect of price of energy on energy consumption, employment, and economic growth needs to be thoroughly examined also.

Solar energy technologies will become increasingly important over the next decade. Also, energy policy in the United States will undergo several stages as we shift from energy dependence to conservation.

The balance between economic stability and economic efficiency can also be viewed as a balance between diversity and simplicity. Diversity within a society implies a good deal of stability. A diverse economy also implies many individual firms or agencies and small units within a large society.

Simplicity on the other hand implies a small number of large units, making possible economies of scale (where they exist). This system tends toward centralization in contrast to the decentralized economy of the diverse society. The diverse society depends on windmills, solar collectors, geothermal plants, and small electric generators, while the centralized society depends on power stations providing upwards of 1000 megawatts. Appropriate energy technologies should be relatively non-polluting, cheap, and labor intensive, non-exploitative of natural resources, compatible with local cultures, functional, and non-alienating. These may be the goals of the United States' energy policy of the next several decades.

An Introduction to Energy

Paul Craig

This last session of the Conference on Appropriate Technology is devoted to energy. When one is dealing with a complex society, it is very hard to find organizing principles. The society is so interconnected and interrelated that it is often hard to determine what the main driving forces are. For our society at this particular period of history there is perhaps more confusion among all of us than has existed in previous years. At least one of the major driving forces is known, however: energy. The fact that we are at the peak now of the fossil fuel era in the history of mankind is becoming abundantly clear to all of us. It should have become clear years ago. As has been pointed out a number of times, M. King Hubbert made it clear years ago but there are long time periods required for change in any society. We are just approaching the point in the United States of recognition of need for change in our patterns of energy use. I am optimistic that within the next few years we will see major changes occurring. That is, we will move from the talking stage to the action stage. Within the federal government (especially the Congress), there has been remarkably slow recognition of the importance of finding ways to modify our energy use. Bureaucratic institutions move very, very slowly. But they are beginning to move. I think the recent papers by Amory Lovins have stimulated an enormous amount of interest throughout the nation, and, in particular, at the Washington level.* It seems to me that it is very

*Amory Lovins, "The Road Not Taken," *Foreign Affairs*, Fall, 1976; and "Scale, Centralization, and Electrification in Energy Systems," paper presented at Oak Ridge Associated Universities Symposium, Future Strategies for Energy Development, October 20–21, 1976; to be published.

likely that within the next few years, especially with the new administration in place, one is going to see energy conservation activities take on an importance they have not taken in the past.

One of the concepts which is likely to become important in the next few years is a principle enunciated by Professor Kenneth Boulding of the University of Colorado (the first person to point out that we live in a finite world, a spaceship earth).* About twenty-five years ago he wrote a beautiful paper in which he pointed out that from the point of view of human welfare it is not the flow of goods through this society which is important, but rather stocks of goods. Gross national product is a measurement of flow, and does not necessarily have very much to do with well-being. For example, it is not the building of a house which makes you comfortable, it is having the house to live in. It is not the manufacturing of shoes, it is having shoes. It is not the manufacturing of phonograph records, it is having the records to listen to. And so on through the entire society. As we increasingly move into a period in which the fossil fuels will be replaced by renewable resources, we shall have to recognize the importance of stocks of goods as contributing to human well-being, while flows are important for replacement purposes and for expansion, but not as ends in themselves.

Finally, I want to make one remark about the University of California and appropriate technologies. The University of California is a land-grant college. Built into the charter is the statement that not only is the University responsible for the development of knowledge, but it also has an obligation to assure communication of knowledge. I expect that in the next few years we are going to see the University playing an increasingly active role in application of appropriate technologies in California. I do not really think at this point we know just which technologies are appropriate for California. We all have our own ideas. There is going to be a revolutionary process to figure this out. The University of California will be developing courses of all sorts—conventional and specialized seminars, and workshops that go for just a day or two. I hope those of you who are interested in more training in appropriate technologies will contact those of us associated with the University and

*K. E. Boulding, "Income or Welfare," *Review of Economic Studies,* vol. 17, no. 77 (1949).

indicate just what sorts of programs you would like to see offered. After all, the only way in which good courses can be developed is if there is an indication as to what you want. Leadership ought to come from you even more than from the University.

Energy, Self-Reliance, and Self-Sufficiency

Steven Slaby

SMOG

I believe in the *vengeance* of history,
  and the *vengeance* of heaven for depravity
Possibly smog is an ectoplasm
Descended for vengeance on the world's baseness.

Darkness is coming,
  Darkness!
It reeks of deepest hell.
Those who can breathe this stench
  Are *not* worth keeping alive!

When the world is a *cadaver*
  A cesspool of fog and chaos
It is a *sign of quality*
  To *sink and drown.*

False ideas,
False morality
Fuming so many years
have soiled
the sky.

Listen!
It is easy to lose your breath on a precipice,
But breathe deeply
  Breathe deeply!
Give it a try!
  Inhale altogether!

You will see—
only inhale,
and the phantom smog
by your breathing will be swept from the sky!

And I felt the epoch
Standing still awaiting,
Like a revolution of the universe,
Our common deep breath.

I've quoted the poem "Smog" by the Russian poet Yevtushenko which appears in one of the collections called *Stolen Apples.**

My immediate remarks now are introductory in nature. There is no energy crisis in the United States. What we are experiencing is a crisis in values based on our perverted order of priorities. We cannot understand the present energy scene without relating what is happening now to the past. The history of politics, oil, and energy is rampant with distortions, lies, conniving, conspiracy, feigned shortages to drive up prices, surpluses to drive out independents, price fixing, bribery, the buying of politicians. Examine the book entitled *The Politics of Oil* by Robert Engler. There's a new book called *The Seven Sisters,* by Anthony Sampson, which documents the history and the politics of oil. It's the same old game. Profits at any price, including the well-being of us as individuals and as a nation. There's no scarcity of oil at the present time. Oil is beginning to flow into the West Coast at such a rate that there isn't enough space for tanks to store it. This is coming from the Alaska pipeline. I get this information from a variety of sources. For one, my brother is a vice president of an oil company. Occasional leaks to the media by the oil industry indicate that moves are being made to sell Alaska oil to Japan. This was predicted by the critics of the Alaska pipeline long before it was built.

Recently, former Secretary of Commerce Elliott Richardson just a day or two before the demise of the Ford administration approved a loan of three-quarter billion dollars, I repeat, three-quarter billion dollars, to General Dynamics, to build two or three tankers to carry liquefied gas from Indonesia to Japan. And this at a time of our reported gas shortage here at home. Now I don't want to be mis-

**Stolen Apples: Poetry by Yevgeny Yevtushenko,* with English adaptation by James Dickey et al., Garden City, N.Y.: Doubleday, 1971.

understood. Just because I'm convinced that there's no real shortage of fossil fuels at this time, it in no way implies that we should not look at our social-economic system with a critical eye to eliminating the immensely wasteful base on which it precariously rests and at great social, economic, and human cost to many people in this country and throughout the world. What I am suggesting, in my presentation here, is that what we need is a major restructuring of our value system and, therefore, our so-called economic system. I did my graduate work in economics, and when I got through I learned that in the present context of economics, economics itself has lost its meaning. The word "economic" has its essence removed and replaced with the concept of profit, which has little relation to being economic. To be economic means to be *judicious* in the use of limited resources. It is clear that the system under which we operate is not economic; it is just the opposite. It is wasteful not only of limited natural resources but abundant human resources including men and women. Waste is what makes the system work and it is the poor people of our country and the world which pay for this waste by living in poverty and by being discriminated against.

The major cause of the so-called crisis that we are in is a crisis in values. The most wasteful activity in which we are perennially involved is our obsession with military and weapons technology, which has cost the United States alone over 1.7 trillion dollars since 1945. Now if we convert this to barrels of oil, we might get an appreciation of the order of magnitude of this waste, and all this expenditure is done safely in the name of national security. I haven't done that conversion; it's a difficult job. It would be an interesting student thesis. Wouldn't it be interesting, for example, to have a couple of students work on the project of determining how many mega-BTU's are required to produce one army tank; all the way from getting the iron ore out of the ground to converting it into iron, steel rolling it, forming it, fabricating it, delivering it. We should calculate all the energy that goes into that one item and then convert that into barrels of oil. One-half of the engineers in the world today, for example, are working on weapons or weapons-related technologies. This cost is being borne by the poor and the working class.

As an example, let's consider housing. One-sixth of the housing in these great United States of America is substandard, in a country where we have the highest and most beautiful technology available that mankind has ever seen. Fifty percent of the housing in Trenton, New Jersey, is substandard. This is the same for Newark, New Jersey; New York; Detroit; New Jersey is also known as cancer alley since we have the highest rate of cancer too. Unemployment in New Jersey has consistently been, for the last few years, over 11.3 percent. Unemployment among employable teenagers in Newark is over 40 percent. There are over 17 million children in the United States who live in such dire economic straits that they have no chance to live fulfilling lives. The suicide rate among teenage native Americans on the reservations is double that of the national average. They have no will to live and no place to go. I relate these examples to the energy and technology issue because we seem to have an endless supply of money and resources to develop instruments of death while there always seems to be a shortage of money and resources for improving the quality of life for our poor brothers and sisters. Can we deal with the problems I have mentioned by tinkering with concepts such as centralization or decentralization of energy systems, or so-called scales of economy or government regulations? Centralization of energy systems and institutions in general has led to massive decentralized waste in the form of blind, mindless consumerism. It has led to a push-button, instant-gratification mentality where we are conditioned to expect almost orgiastic satisfaction each time we buy a new gadget, in spite of the fact that the premise of this experience nevr meets our expectations. I picked up a little thing the other day to discuss with a class. It is symbolic of what is going on. It's a butane lighter, it's disposable, and in the same package was a disposable razor. It's called a Cricket and the heading on it was, "Good news!" The suggested price was $1.79; crossed out was $1.49; and then a little sticker on the package: sale price was 77 cents. So that gives you a little idea of the kind of game that's being played. But this is illustrative of the disposable mentality! I bought it and I'll use and then I'll dispose of it, but I won't buy it again. I'll throw it away in front of my class and see if they'll calculate the energy loss.

The centralization of energy resources and systems has put us as individuals and as a nation at the mercy of large corporate interests. And in the process we have lost our self-reliance and self-sufficiency and have become alienated from each other. Furthermore, we have become subservient to centralized political power centers where the illusion of freedom is perpetrated on us as the democratic processes are corrupted and citizen participation in decisions which affect our lives is orchestrated by the politicians and those in power in such a way as only to perpetuate their continuance in office. In actuality, we have no role of any real significance in controlling the affairs of the system and the state as it is presently constituted. We are at the mercy of bigness—big business, big government, big institutions. In other words, we are at the mercy of what some people call the establishment. It is evident that large-scale entities have resulted in effects that are dehumanizing and continue to have a depersonalizing impact on our lives. People become secondary in value to the workings of large-scale systems. Racism and sexism continue to flourish since the system is so large and impersonal that individuals are detached further and further from each other as well as from their social responsibilities to each other. The impact on individual and societal values is such as to reflect the values of big business which result in a bureaucratic mentality where the bureaucracy becomes a cancer growing on the backs of the poor and the working class. Now what is interesting is that there is one common brotherhood in this world that works beautifully, and that's the brotherhood of the bureaucrats. Whether it is in a communist nation or a capitalist nation, when you meet the bureaucrats and have to work with them, you cannot distinguish one from another. The justification for largeness is the illusion of efficiency and economy of scale. The illusion works because as the corporations get larger, prices for their products go up and not down, and we are told that this is because of inflation, as if inflation were caused by some divine intervention in the affairs of humankind. Tying centralization to large scale with government regulation, we arrive at the point of truth. Big business is big government. Regulatory agencies are closely allied with those they are supposed to regulate. Examples of this can be seen in the airline industry, the nuclear power industry, the oil and gas industry. What are the possibilities and realities of transforming the United States' energy system to an intermediate technology base? The possibilities are limited, in my own opinion, since those who have political power, the political power elites, will never give up their power without a struggle. This is an historical fact. I can't recall any point in history where power has been given up freely. If intermediate technology begins to threaten their power, the elites will co-opt and infiltrate this movement and destroy its effectiveness and its integrity, as they did the black liberation movements and the anti–Vietnam war movement as well as left political groups. The record shows clearly how the CIA infiltrated the National Student Association and how the FBI infiltrated various political groups. If "small is beautiful" is to succeed, it must retain its integrity as a concept in terms of a people's technology and a people's economics and a people's political movement. It cannot permit itself to be used as a cover for covert activities or clandestine government agencies. And it should retain its independence from government agencies by refusing to accept their financial support and sponsorship. If we permit the opposite to happen, then the small-is-beautiful concept will only be a placebo which makes you feel good when you are really sick.

Intermediate or appropriate technology, as Dr. Schumacher stated, involves more than just hardware. It includes all elements of the human spirit and therefore all who are involved in it must be part of a developing community which can ultimately withstand the onslaught of the establishment because after all, if small is beautiful is a serious concept, it must result in a radical restructuring of our value systems to create a humane order of priorities with appropriate energy systems; in other words, it will ultimately result in a radical restructuring of our entire economic and social system and if you and I are truly concerned about the state of our world and its future, let's begin this long-term process by transforming ourselves into men and women who will each be willing to choose a role in what in actuality is a political movement of humane dimensions. And this kind of political movement, if you are serious about it, requires personal, communal, and national sacrifices, which will transform our life style from one based on death to one of living our lives to the fullest. I would like to end my presentation with another Yevtushenko poem. This is entitled "Monologue of a Broadway Actress."

MONOLOGUE OF A BROADWAY ACTRESS

Said an actress from Broadway
  devastated like old Troy:
There are simply *no more roles.*

No role
    to extract from me all my tears,

No role
    to turn me inside out.

From this life, *really,*
    One must flee to the *desert,*
There are simply *no roles anymore!*

Broadway *blazes*
like a hot computer

But, believe me, *there's no role—*
    Not one role
Amidst hundreds of parts.

Honestly, we are *drowning* in rolelessness. . . .

Where are the great writers! *Where!*
The poor classics have broken out in sweat,
    Like a team of tumblers whose act is too long,

But *what* do they know
    About *Hiroshima,*
Abour the *Murder* of the *Six Million,*
About all our pain?

Is it really all so *inexpressible?*
Not one role!

It's like being without a compass.
You know how *dreadful* the world is
When it builds inside you,
    Builds up and builds up,
And there's *absolutely* no way *out* for it.

Oh yes,
    There are road companies.
For that matter,
There are T.V. serials.
But the *roles* have been *removed.*
They put you off with *bit parts.*

I *drink.* Oh I know it's weak of me,
but what can you do, when there are *no more people,*
No more roles?

Somewhere a worker is drinking,
    From a glass—*black* with greasy fingerprints.

He has no role!

And the farmer is drinking,
    Bellowing like a mule *because he's impotent,*
He has no role!

A *sixteen-year-old child*
    Is *stabbed* with a *switch blade* by his friends
    Because they have *nothing* better to do. . . .

There are no roles!

Without *some* sort of *role, life*
    Is simply slow rot.

In the *womb,* we are all *geniuses.*
But *potential geniuses* become *imbeciles*—
Without a role to play.

Without *demanding anyone's blood*
I
Do demand
    A role!

California's Approach to Energy

Emilio Varanini

First I'd like to acknowledge that there are some state officials who are interested in these programs and have taken an interest and lead in providing what protection the environment has received. In the front row, and I'd like him to stand, is Senator John Dunlop, who led the charge on the coastal preservation protection and various other ecological measures. I'm a bureaucrat. I left my pencil, tie, and short-sleeved white shirt back at the office. I also happen to work for you, and you get in some sense what you deserve. It's your or my vigilance or lack thereof which results in many of the problems which were just enunciated by the last speaker. Several days ago, to indicate the diversity of my commission, two of the commissioners spent several hours of the legislature's time indicating that we were dedicated to no growth. During that time I would venture to say, other than representatives of utilities and oil companies, there were probably not two real citizens in the audience. So it's that kind of thing that's going on in the trenches in public policy, and I'm glad to be here today.

What I'm going to try to talk about is the commissions' biennial report. What we were supposed to do was to report to California every two years on a whole variety of issues. The one I'm most concerned with is future policy. Where are we and where are we going from here, if anywhere? We also try to lay out for the public and the legislature certain ideas and concepts of how we can get from here to there without tearing up everything in the process. What we've tried to do is lay it out in the sense of where are we and what kinds of policy collision one can see. We tried to keep this thing a little bit self-critical internally, and so we tried to start out with a bit of levity by showing what our perspective is. And that is one of the problems, an attempt to illustrate the broad-nature view of energy and our policy. Someone has asked whether it ought to be called nebulous, and I guess that some have that point of view. But basically even the Energy Commission can't reverse the ever-increasing entropy in the system as much as we'd like to.

We think the critical issue in California and the United States is really attempting to deal with uncertainty. The graph shown in

FIGURE 1

VARIOUS PROJECTIONS OF ANNUAL DEMAND FOR ELECTRICAL ENERGY

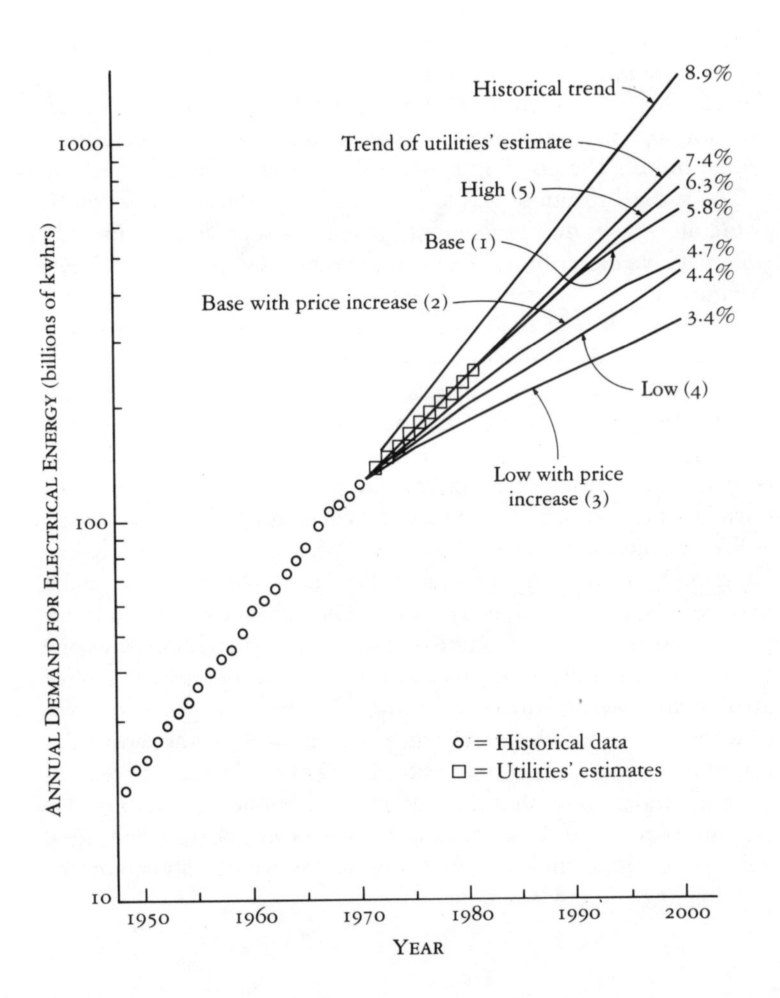

figure 1 represents various projections made in 1972 by the Rand Corporation for the California State Legislature in terms of projections of energy through the year 2000. You can note the case of 8.9 percent growth rate per year and the low growth case at 3.4 percent. It's interesting that for 1975, the lowest imagined forecast was 14 billion kilowatthours too high. That's how good we are at predicting the future. That's a three-year span, but what happened in the meantime? Well, we had an oil embargo on the way to our summer vacation.

Americans had a traumatic experience, other than governmental. They lined up for gas. Even Los Angeles, where in the sixties there had been tremendous price wars with the prices constantly going down, oil companies couldn't get rid of gasoline fast enough. Then we had the federal government's next response, Project Independence. This was a trauma-relief program. The Nixon and Ford administrations noticed that we became and have become increasingly reliant on oil. The graph in figure 2 shows some of that oil growth.

The federal response was to get-rid of the vulnerability of oil by going all-out on domestic energy resources such as coal. In addition, massive electrification was programmed and planned as shown in figure 3.

However, Project Independence ran into tremendous suspicion, particularly in the Congress, which has some expertise in the area. The oil company profits increased dramatically and some thought that energy prices might be manipulated by the oil companies. The administration responded to that by proposing price deregulation of gas, also opposed in Congress, and privatization of nuclear fuel enrichment, also opposed in Congress and stopped by one vote.

We are going to have a sequence of events, so anything can happen. We think basically what we have in Washington now, and it may be relieved in a bit, is a policy deadlock between what was our policy and that of the Congress. The overreaching issue, the one that's used in some sense to emotionalize the process, is that of growth and the limitations to growth. There are those who advocate that we should limit growth, particularly economic growth in use of non-renewable resources for energy for a number of reasons but the primary ones that they see are catastrophe. Civilization may outrun its resource base, making poor nations poorer if industrial nations continue their resource use and as well a catastrophe by overloading the environment.

FIGURE 2

UNITED STATES OIL CONSUMPTION

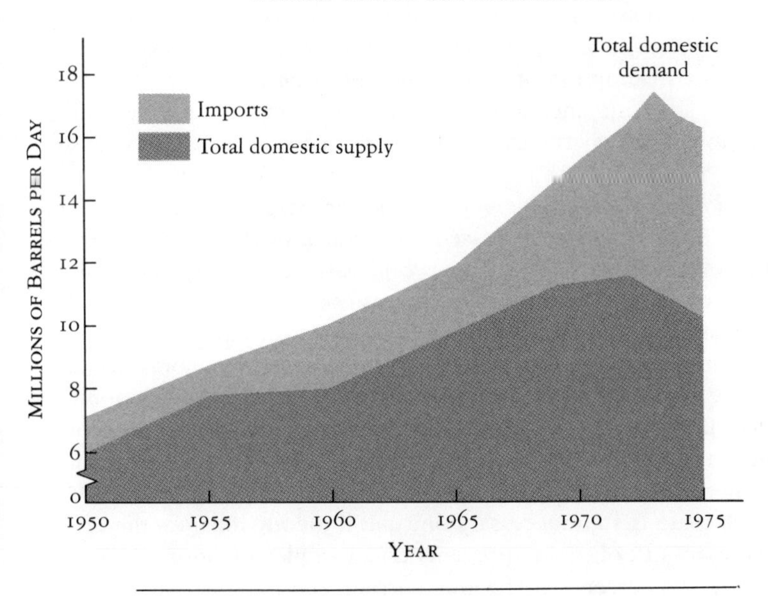

FIGURE 3

PROJECTED GROWTH OF ELECTRICAL GENERATING CAPACITY

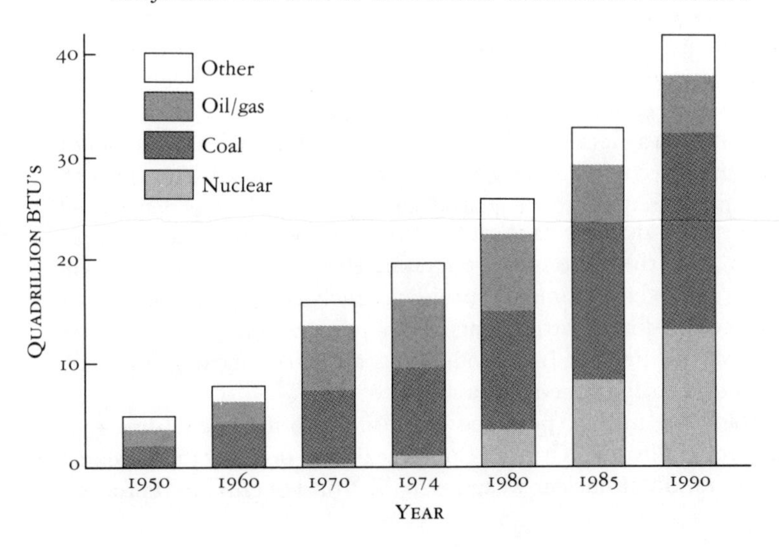

The proponents of limiting growth point to regional catastrophes such as the drought in the Sahel. Some attribute this to the intensive use of fossil fuels and their implications on the atmosphere. They see such catastrophes occur even more frequently as a result of man's consumption of non-renewable resources.

Many advocate increasing economic growth in resource consumption. They say the use of energy resources goes hand-in-hand with economic growth and that economic growth meets the basic aspirations of most people in the world for increased well-being.

Typical advocates of increased economic growth hold that limiting growth will only exacerbate world national poverty. The only way to overcome poverty is to increase the size of the economic pie. They indicate that existing institutions will not shift and redistribute wealth and income and that a steady-state economy would relegate the poor to remain poor. We think that this dichotomy concerning growth is a false one. We know that some level of energy use is required for increasing well-being. But certainly not the energy use that we see here in California. So in many ways this growth issue is an unnecessary one and it merely inflames the discussion about policy and prevents policy implementation.

What we perceive to be our current energy situation is that basically our natural gas deliveries are declining and we are using a lot of electricity and a lot of foreign oil.

Who's going to plan our future in energy? We believe that if current business-as-usual continues that the energy corporations will. What are those plans? They perceived increased demand for energy, increased oil use, and depleting gas reserves.

California's electric utilities basically plan for nuclear baseload to take the place of oil and meet the increase in demand. They plan 28 new nuclear units to be placed and operating between now and 1995. Why do they take this path? They perceive it as the least costly of all the options for generating electricity. They think it will solve major environmental problems such as air pollution, that it replaces oil-fire electric plants. They perceive that nuclear power follows the Project Independence goals as the most appropriate transition to the breeder economy.

Many see nuclear power as a transition to the twenty-first century for California. Critics of nuclear power point to the potential proliferation of nuclear weapons in the world because of reprocessing of nuclear fuels. We heard of this in the presidential debates and in subsequent pronouncements.

Critics of nuclear power also point to the safety problems such as the incident that occurred at TVA's Browns Ferry, Alabama, nuclear plant prior to the fire. They also see massive problems with enrichment capacity, waste disposal, with the economics and the safety of reprocessing.

So what are the prospects of nuclear plants in California? We find them tremendously uncertain. We find the economics are uncertain. Nuclear Regulatory Commission regulations are uncertain. We find that our own commission presents a major uncertainty to those plants, in implementing nuclear bills that were passed by the legislature.

A similar situation faces the gas utilities. There's a scramble on for gas supplies, and gas utilities are planning a number of large, risky projects to try to bring new gas supplies to the state.

On the North Slope of Alaska, the project favored by the Federal Power Commission staff to bring the gas to the lower 48 states is a pipeline 4,500 miles long that would cost somewhere between ten and twenty billion dollars. There are major financial problems in engineering, and technical problems with these major gas projects, not to mention regulatory uncertainties.

The gas utilities also plan to bring liquefied natural gas to California in huge tankers, which pose a risk of potential disaster should there be a major LNG spill. What we see in the wake of any potential disaster is regulatory uncertainties, including higher costs and potential complete shut-down of entire projects. So reliance on these big projects is terrifically risky. The gas utilities also hope for coal gasification plants which cost more than a billion dollars each. And because it's something of a new technology, the gas utilities would require loan guarantees or some other mechanism to share the financial risk with you. What are their prospects? They are tremendously risky and uncertain. None of the new major gas-supply projects may come to pass.

The oil situation on the North Slope of Alaska is substantially different but illustrates a number of themes. The building of the Alaskan pipeline was first estimated at costing about $900 million. The estimates of what it costs now are approximately $8 billion. This illustrates our theme of uncertainty and potential for disaster.

In the early 1970's the oil companies assured the Nixon administration that all the oil would be consumed on the West Coast. We now know that the construction of this pipeline which carries oil to the Pacific Ocean and to the West Coast, where all of it is not likely to be needed, poses serious problems about how to transport the oil to other parts of the country, so that a large inflexible project has now got us into environmental and regulatory trouble.

To illustrate the theme in the report, the environmental impact statements of these oil projects indicate that another major oil spill is a near certainty. The Santa Barbara oil spill, a major disaster in 1969, caused a regulatory backlash. It's called good government if you are on the other side of the issue. It caused a new imposition in the cost of offshore drilling, it caused the shutting down of leased offshore oil areas, and was a major motivating factor behind the National Environmental Policy Act which went on to require EIS's from federal agencies. So what we see is that when you get into a major risky energy project it could cause a disaster. Environmental risks, additional costs, and even potentially losing a source of supply may occur. Increased oil production and transportation also comes into conflict with our permanent California Coastal Commission.

Of course increased oil use in oil transportation and production causes air pollution, which comes into conflict with the California Air Resources Board Control District, Environmental Protection Agency, and rules and laws designed, hopefully at least, to protect your lungs. So what are the prospects? Again, uncertainty. Conflicts, environmental groups fighting with energy companies, regulatory problems with projects, financial, technological, and engineering problems. What do we do? Oddly enough, we see a lack of consensus.

Small is beautiful. This point of view holds that the remaining energy should be used in a more dispersed, softer, and thermodynamically appropriate technology. The soft and human-scale technologies do away with many of the risks we mentioned earlier. They do away with financial risk: the projects are small. The risks of disaster with light and wind are eased.

Some people indicate there are even psychological benefits from living in a log cabin. Critics of the point of view hold that small is beautiful is alien.

Yet what about the aspirations of most Californians? They say if small is beautiful is used as a base for planning, people's aspirations of material well-being remain the same. You run the risk of social unrest and even revolution.

Now we have the second point of view, which we've called somewhat euphemistically the *Greeleyan* view. This second point of view holds that government should return to the days of promoting business developments such as when government gave land to railroads. Proponents of this point of view hold that many of the risks mentioned earlier, particularly the financing risk, are caused by government regulation—that the huge number of permits required prevent business from providing the goods and services that American people want. And therefore, such regulations should be simplified, decreased, clarified, and stop changing so frequently. Proponents of this point of view also hold that these projects provide substantial net benefits and therefore the share in financial risk with government is warranted.

The critics of the industry point of view maintain that the risks are real and they should be substantially regulated. They also hold that many of the processes by which resources are developed are a giveaway, a rip-off to the consumer, and that big business shouldn't need loan guarantees, subsidies, and other such financial help from the people.

The third point of view is what we call *Periclean.* What is needed is bigger and better comprehensive governmental planning. The comprehensive planning would signal to environmental groups and developers where development could take place, and coordinate it rather than having single-purpose regulatory agencies. This way of thinking perceives order out of chaos through the government. Critics of this point of view, and there are many, point to such projects as the Bay Area Rapid Transit System as examples of governmental failure.

It's causing such unforeseen environmental consequences such as sprawl resulting in cost overruns similar to those experienced by energy corporations. They hold that developing consensus is impossible and increased comprehensive planning would create only another layer of government that would cost more with no benefit. And then there's the final, and sort of a ultimate point of view. This is the so-called *technological optimist fix.*

This is at least our attempt to quantify the philosophy of what we propose as eclectic-hedging strategy, until we are informed that you the people don't understand what any of those words mean. So you can read them. For those of you who can't I'll tell you afterwards. We'll keep it a secret. Basically we are attempting to avoid dependence on fuel technology for resources, and to minimize dependence on large projects. We are looking for some flexibility. We don't want to lock the state into long-term projects, we want the state to take advantage of breakthroughs. There were some interesting comments about modern computer components, and we think if one were to compare the cost of Univac with the cost of a hand calculator that it might be in the same general range as talking about the difference and projection of solar cells now and solar cells if we really put our mind to it. If we're locked into amortizing billions of dollars in burner and breeding reactors, that new, small technology will have to wait. If you read the technical press, these types of things in terms of cost amortizations tend to play a fairly important role in when we're going to see our next toy for our consumptive heart's desire. Do what we know how to do; moderate size, known fuels and technologies; minimize regulatory uncertainty; and conserve energy.

I think there's a tremendous potential for energy conservation in California. We think you'll find a way.

If we have a plan with all the paraphernalia and all the other elements that one runs into, these are some of the criteria: retrofitting insulation, efficient heating systems, outreach services, commercial building delamping, industrial cogeneration of waste heat recovery, load management, and marginal cost pricing, are some.

We think there are appropriate uses of certain types of new technology for electric generation and alternatives which should be looked at. We will be proposing a package of measures to accelerate implementation of solar. One of the things, of course, is for you to want it, but secondly, there are many institutional barriers to solar which are designed to keep the technology off the market, and we're going to try to set those forth to help eliminate the statutory barriers by aiding the legislature in pointing where they are. The first example is that cost comparisons between solar and competing technologies are laughable. Basically there is a series of barriers in the construction code in California which are inane, thoughtless, ancient, and make no sense. We have laws that basically don't guarantee your rights to the sun after you put the collector in. We're subsidizing a lot of other things you are receiving that you don't want, so it might be possible to subsidize something that you might want.

We suggest a clean fuel strategy that is the use of cleaner fossil fuels such as methanol and liquified petroleum gas. We'll suggest some innovative financing arrangements to the legislature in terms of these alternative technologies.

We're attacked very heavily when we propose government involvement in financing, so we attempted to point out to the energy companies that they had come to the legislature and asked for legislation to put through little billion sweetheart contracts, and we thought if it was good enough for them, it would be good enough for us.

These are some of the advantages we see in the philosophy, and again, we're not trying to get into detail here, we are trying to minimize risks in terms of disaster, regulatory obstacles, surprise cost, supply interruption. We want to know what these things are going to cost. The philosophy doesn't require a total and absolute consensus. The obstacle is that we have got to do something, and try to get out in front for a change. At this point in the talk I used to say that such an incremental, eclectic strategy didn't require consensus and that it was practical and implementable. But we all live and learn.

Some aspects of the commission's 1977 biennial report have been heavily criticized. I think the real bottom line of this is that government is going to have to take some form or role, otherwise we will run into a situation such as we did with the Alaskan pipeline, where we were promised by Secretary Morton it would be done in an ecologically sound manner and for that we gave up the National Environmental Policy Act and we gave up the ability to sue. We gave up a whole series of rights as citizens in that case. And now we have a pipeline that might leak, we have a pipeline that might not work, we have a receiving point where the ships may not be able to enter, we have ships without double hulls, without twin rudders, and we have ex-Secretary Morton to pursue, wherever he is.

Fortunately or unfortunately, government cannot get out of the way, and whether one believes it or doesn't, one is going to have to work or at least attempt to work through it or change it in order to resolve the issues that I've set forth. This report will be available in about a week from the commission, it's our biennial report overview, and it's an attempt to try to give some guidance to you about policies in California, actions you can take yourself, actions you can support, and an opportunity for you to tell us to go take it and do whatever we want with it if that's your view of this report.

The Potential for Energy Conservation through Local Initiative: A Case History

Jonathan Hammond

After more than two years of research, planning, and public debate, the city of Davis, California, adopted in November of 1975 the nation's first comprehensive, energy-conserving building code tailored to the needs of a specific microclimate. Work on the code began in the spring of 1972, when a newly elected city council, committed to controlled growth and environmental quality, commissioned a study of ways to save energy in Davis homes and apartments. The council's action was prompted by the citizen advisory group's report on energy consumption patterns in Davis. The report found that Davis residents use one-third more electricity per capita than that used in the average American town. Apparently, the poor adaptation of Davis homes to the relatively mild local climate accounted for their surprisingly high rate of energy use. In an attempt to correct this situation, the city hired the research team of Marshall Hunt, Loren Neubauer, Richard Cramer, and the author to develop building code and neighborhood planning policies aimed at conserving energy. The city and the University of California, Davis, jointly funded the project. The results of our research and the energy conservation measures which evolved from it are the subject of this article.

Developing a residential building code to conserve energy seemed at first a straightforward problem of fitting an appropriate building technology to a given climate. However, our dealings with Davis building designers and homeowners soon made it clear that cultural as well as physical factors prevented energy-efficient housing. Our first task, as a result, was educational. We found that few people knew how to operate their homes efficiently, how to choose well-designed housing, or how to improve their homes' thermal performance. Our preliminary work was designed to address these areas by investigating the performance of Davis buildings and presenting the results in a way which would help local citizens understand how climate affects the buildings they live and work in.

There is a well-developed body of knowledge and comprehensive set of techniques available for calculating the heating and cooling needs of buildings. Yet these techniques, useful as they are to the engineer, are too complex to be used by the average person. Most people know more about the thermodynamics of nuclear reactors than about how their own home works. We therefore decided to study real buildings for the following reasons: (1) the testing of actual buildings would emphasize how real buildings worked, rather than the efficiency of our research techniques and models; and (2) focusing on the buildings would allow citizens to see in a familiar context how fundamental design features worked in relation to the local climate.

Davis is located in California's great Central Valley. The valley has cool, moist winters with temperatures averaging around 45° F with minimum temperatures seldom colder than 25° F. The average rainfall is around 16 inches per year, occurring almost entirely in the winter. Even in the winter, the sun shines more than fifty percent of the time, providing a good opportunity for solar heating.* The summers are warm with daily maximums averaging around 95° F; 100-degree temperatures are not uncommon. These extreme daytime highs are, however, considerably moderated by sea breezes coming through the Golden Gate and entering the valley through a gap in the Coast Range at the Carquinez Straits. These sea breezes typically arise in the late afternoon or early evening and rapidly cool the area, creating an average summer night-time low of 53° F. With this kind of natural diurnal cooling it seems ridiculous that some Davis homeowners and apartment dwellers pay in excess of $40 monthly for air conditioning. The problem of cooling houses is particularly important since the demand for residential air conditioning occurs during California's peak electrical demand period.

---

*University of California Agriculture Extension Service, *The Climate of Yolo County*, Davis: University of California, Davis, 1971.

Typical Davis apartment used for thermal tests

We wanted to see how typical Davis buildings responded to the valley climate and how much energy they were using. Our research would demonstrate how differences in construction, materials, orientation, windows, color, insulation, roof overhangs, and vegetative shading affect thermal performance. The following discussion summarizes the findings included in a report prepared for the city as background for the proposed code.*

In the summer of 1973, temperature-measuring equipment was installed in finished, but unoccupied apartments; apartments provided especially useful comparisons, since some had identical units facing the four cardinal directions. The orientation, window size and location, and roof overhangs of such units could be easily compared to their interior temperatures, levels of comfort, and energy use. Three different apartment complexes were analyzed, all well built by conventional standards.

---

*Jonathan Hammond, Marshall Hunt, Richard Cramer, and Loren Neubauer, *A Strategy for Energy Conservation,* Winters, California: Living Systems (Rt. 1, Box 170), 1974.

In July and August, interior air temperatures in unoccupied units were monitored. The apartments were two stories high with individual flats oriented to the exact cardinal directions. It was found that units with north-south exposures were much cooler than east-west units. The coolest units faced north-south on the ground floor: they had a mean maximum of only 75° F, perfectly comfortable in hot summer weather. The hottest apartments faced east-west on the top floor. At 99° F, they were 24° hotter than the coolest apartments, and impossible for human habitation without expensive air conditioning (see figure 1).

FIGURE 1

AVERAGE APARTMENT SUMMER TEMPERATURES

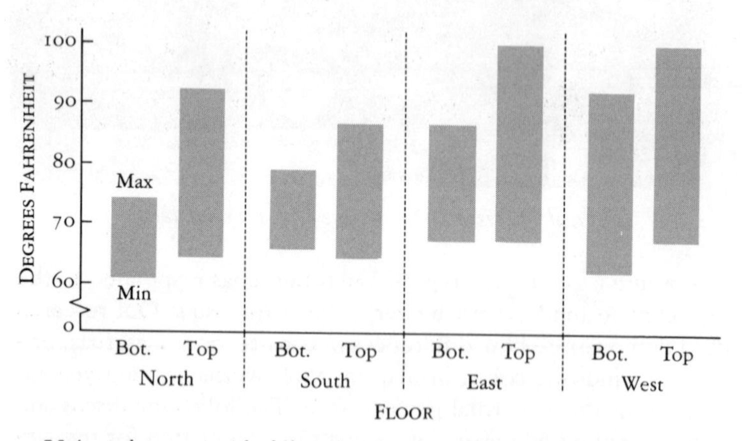

Using data provided by the Davis office of the Pacific Gas & Electric Company, twenty-seven occupied apartments in another part of the same complex were analyzed to see if actual energy use correlated with our findings. Figure 2 shows the results obtained by subtracting the average winter month's electrical use from the average summer month's use. We assumed this would isolate the energy used by the air conditioner in these units from normal electrical usage. The top-floor apartments consumed twice as much energy as the middle-floor units, and nearly three times that of the bottom-floor ones. Inadequate roof design accounts for the high energy use of the top-floor units. The units on the east side were worst of all since they not only had inadequately insulated roofs, but also received early morning direct solar heating through their large, east-

FIGURE 2

AIR CONDITIONING ELECTRICAL USAGE
(Average summer month's electrical usage
minus average winter month's usage)

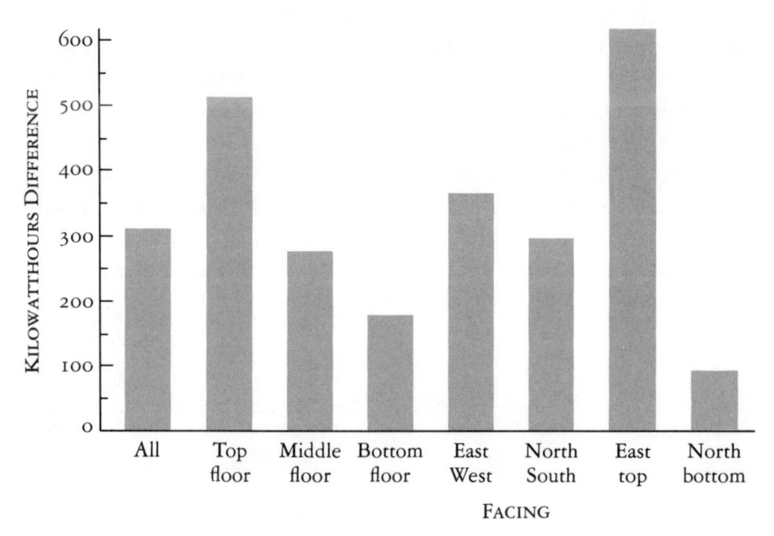

facing windows. In these apartments the air conditioner started early in the morning and ran all day.

Both apartment temperature and energy use demonstrated the critical importance of window orientation. The only really effective means of protecting poorly oriented windows is an exterior shading device. Our tests showed a heat reduction of only one to five degrees when interior drapes were used to shade east- and west-facing windows.

A winter study similar to the one described above was also conducted to show the heating potential of south-facing windows. We found that south-facing apartments with good sun exposure were significantly warmer in winter. Temperature measurements were taken in several unoccupied apartments during sunny, clear, cold days of December 1973 and January and February of 1974. Since apartments facing north, east, and west shared similar maximum and minimum temperatures, they were grouped and averaged for comparison with south-facing apartments and ambient outdoor temperatures. On several occasions, south-facing apartments registered temperatures in the 80's on sunny winter days; the maximum was 87° F. For several days the high temperatures were 24° above ambient, and 17° above north-, east-, or west-facing units. These temperatures occurred in apartments without very large south-facing windows. In comparison, a cubical research room with a completely glass south-facing wall registered an interior maximum 48° F above the maximum ambient temperature.* These tests demonstrated the considerable savings in heating costs and energy use possible in Davis and similar climatic areas through the use of good southern exposures.

Winter natural gas use for space heating in occupied apartments confirmed the value of direct solar heating provided by south-facing windows. Units facing south used less gas than other apartments, as much as thirty percent less than north-facing units. Figure 3 shows average gas use for winter heating.

FIGURE 3

WINTER APARTMENT NATURAL GAS USAGE

(Average winter month's usage minus average summer month's usage)

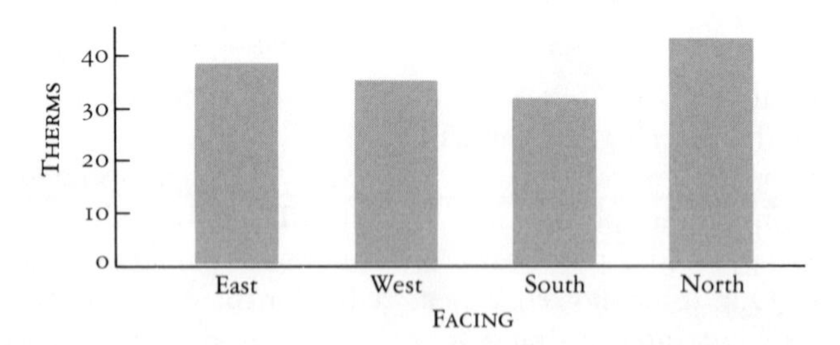

To compare detached houses to apartments, gas and electric use data was obtained from a sample of typical houses. The results indicated that the average Davis apartment uses 25 percent less electricity and 39 percent less gas per square foot than the average Davis detached house. Common walls and reduced outside surface area accounts for the apartments' superior thermal performance.

The energy use of various neighborhoods was also monitored and compared. The oldest part of Davis had the lowest electrical

---

*R. D. Cramer and L. W. Neubauer, "Solar Radiant Gains through Directional Glass Exposure," American Society of Heating, Refrigeration and Air Conditioning Engineers, *ASHRAE Transactions*, vol. 65, no. 59, 1959, p. 499.

use, but surprisingly a relatively high gas use per square foot. We attributed this to the fact that older houses are usually well shaded by mature trees, so they need little air conditioning.

Even though many apartment units and houses in Davis perform very poorly, we found a sufficient number that clearly showed the possibility of achieving good thermal performance with standard building techniques. The conclusions we drew from our study of Davis buildings were inescapable.

1. The basic frame construction techniques almost always used could easily produce buildings reasonably well adapted to Davis climate.
2. Simple design parameters for creating climatically well-adapted buildings were consistently ignored by the local building industry.
3. Existing city policy and citizen review processes did little to encourage better buildings. In fact, in some instances review policies resulted in reduced energy efficiency.
4. If high levels of energy efficiency were to be achieved, Davis needed a building code requiring future dwellings to adapt their design to the needs and opportunities of its climate. In addition, some of the city's neighborhood planning policies would have to be revised to accommodate aspects of the proposed code.

In order to be workable, it was decided that the code would have to meet the following criteria. First, it must be performance-based for the Davis climate. Second, it must be flexible and easily understood. Third, it must not significantly raise construction costs. Fourth, compliance must be possible using standard building technology and building practice. Fifth, it must save significant amounts of energy.

In order to allow for maximum design flexibility, we proposed a code involving two alternatives, between which a builder could choose. Path I is prescriptive, providing a set of rules and possible trade-offs that if followed would result in compliance. Path II sets a minimum performance standard; it describes a calculation technique and specifies a design day for determining the compliance of buildings not conforming to the first path. Path II's performance standard is based on a heat loss or heat gain per day per square foot of floor area. For instance, for a 2,000-square-foot house, the heat loss could not exceed 192 BTU's per square foot per day on the winter design day, and 95 BTU's heat gain on the summer design day (infiltration was not included in these figures). Path II allows the most design flexibility and was designed to encourage innovative solutions in adapting buildings to the local climate. Path I, on the other hand, allows compliance without the use of detailed calculations. We anticipate that more than ninety percent of future buildings will comply with Path I. Some of its more important provisions follow:

1. All wood frame walls must use R-11 insulation, and roofs must use R-19. Some exceptions are heavy-weight walls of concrete or brick. Structures with such walls can use less insulation, if it is placed on the outside, because of the compensating heat storage capacity of the heavy walls which help dampen temperature fluctuations.
2. Overall window area is limited to a percentage of the floor area, specifically 12.5 percent for single-pane glass. The window area can be increased by using double-pane glass or by using south-facing windows which take advantage of winter sun. All but a small percentage of the window area must be shaded in summer. The seasonal geometry of the sun's path resolves the seeming conflict between summer shading and winter sun exposure. Since the sun is high in the summer and low in winter, properly designed overhangs on south-facing windows can provide summer protection while at the same time allowing for winter solar heating.
3. Good natural or artificial cross ventilation must be provided to take advantage of Davis's cool summer evenings for cooling the mass of structures.
4. Light colors must be used on roofs in order to reflect summer heat.

An estimate of the economic impact of these requirements developed with the aid of local builders, indicates that slight increases in design costs, insulation, and shading devices would be offset by long-term savings in utility costs. Our analysis indicates a net monthly savings of approximately five dollars for the owner of a typical 1,500-square-foot home.*

*Hammond et al., *Strategy for Energy Conservation.*

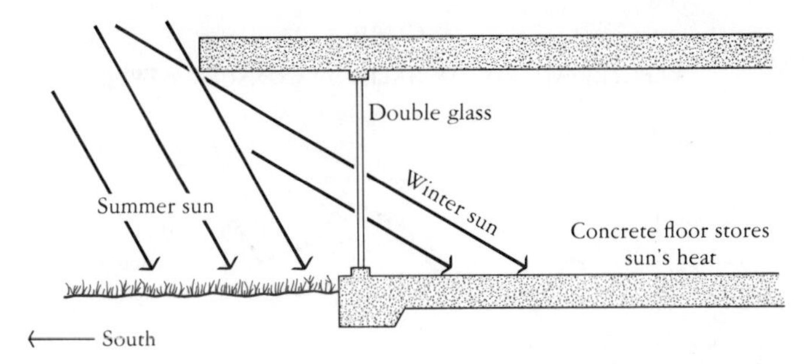

Solar window

A simple case demonstrates some other benefits of saving energy. Every fifty-five square feet of unshaded west-facing window in a typical Central Valley home increases that house's air conditioning need by approximately one ton. Since the peak hour for electrical demand in California corresponds to the demand for residential air conditioning (figure 4), an extra ton of air conditioning (with an energy efficiency ratio of 6) will increase the peak-hour demand by approximately two kilowatts.*

We have a choice: buy an additional ton of air conditioning at an approximate cost of $250 and spend another $2,000 to increase our generating capacity by 2 kw—total cost $2,250; or eliminate the solar radiation entering through the window. Shading the window with a movable insulation system would cost approximately $5 per square foot; a less efficient but adequate system using a movable metal shade screen would cost about $1.50 per square foot. The old standby of a bamboo screen hanging from an eave would probably cost less than 25 cents per square foot. For shading the window the costs run from a high of $275 to a low of $13.75. Even the most expensive shading device costs only one-eighth as much as providing air conditioning to take care of unshaded window impact. The least expensive shading strategy is 169 times less expensive than air conditioning! This is a specific case, yet typical of a situation which repeats itself hundreds of thousands of times in residential and commercial buildings throughout California. The waste of energy and capital involved is obvious.

---

*State of California, Energy Resources Conservation and Development Commission, "Quarterly Fuel and Energy Summary, January–March 1975," September 1975. One ton of air conditioning removes 12,000 BTU's per hour.

FIGURE 4

ELECTRICAL UTILITY SALES BY CONSUMER TYPE

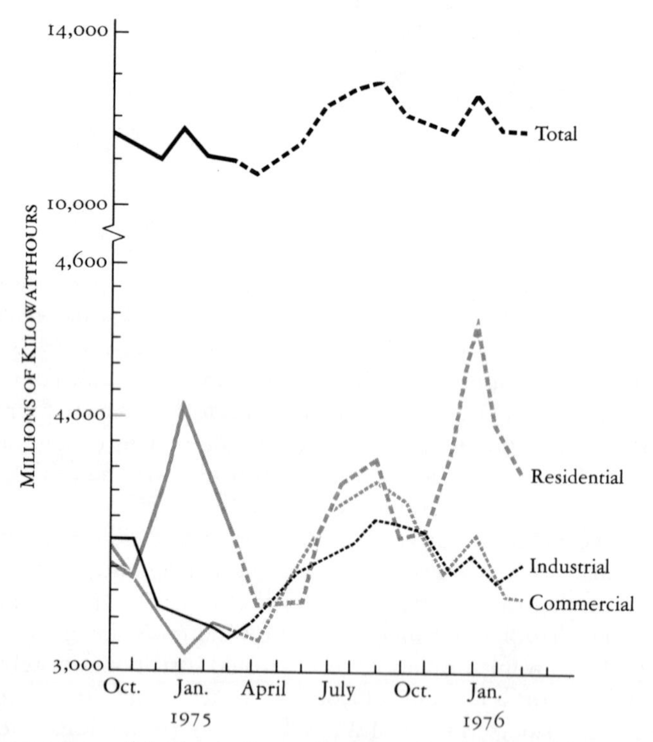

SOURCE: State of California, Energy Resources Conservation and Development Commission, "Quarterly Fuel and Energy Summary, Jan.–March, 1975," September 1975.

NEIGHBORHOOD PLANNING

Neighborhood planning is a very important part of any overall strategy for conserving energy. The following discussion, drawn from some proposed neighborhood planning policies submitted to Davis, suggests some simple ways of saving energy through careful neighborhood design. The proposals emphasize good solar use and complement our work in revising Davis's building code. They are currently being reviewed.

There are a number of elements involved in designing energy-efficient neighborhoods. They include street layout, lot size and shape, and the size, shape, and height of buildings and vegetation. One of the first effects of our work in Davis was that as developers realized that it would be easier to comply with the code on lots with good solar orientation, the street and lot layouts of several proposed subdivisions were voluntarily revised to maximize the number of lots with north-south orientation.

As the price of conventional fuel rises and solar heating techniques become more available, a site with good solar orientation will be more valuable than other sites. To protect the value of such property and to ensure the owner's right of solar use, we have proposed the adoption of a solar rights ordinance. This ordinance would utilize the concept of "envelope zoning." Envelope zoning establishes a three-dimensional shape in which an owner can build without unduly shading a neighbor's property. By requiring envelope zoning for new subdivisions, future legal problems should be avoided.

Another factor critical to allowing maximal solar utilization is the problem of setbacks. Typical subdivision requirements dictate a rigid placement of houses and fences on lots. Greater flexibility to move the buildings around within the confines of an envelope could improve solar potential. For example, our research clearly demonstrated the value of south-facing windows for winter solar heating, and we anticipate that many buildings constructed under the provisions of the new code will have such windows. Owners of these houses should be encouraged to keep their drapes open on clear winter days in order to use their windows' heating potential. Few people will do this, unless they have a fence, a wall, or a screening hedge between themselves and the street. If the privacy screen is too close to the home, it will shade the windows and waste their potential for winter heating. In Davis, as in most towns, subdivision standards typically require that fences or hedges be twenty feet back from the sidewalks. Such requirements not only adversely affect solar heating, but also create acres of manicured, but useless, front yard space.

In the Davis climate, trees are probably the best resource for moderating summer heat. The shading by trees can reduce a building's interior temperature by more than twenty degrees F, while at the same time increasing the comfort of people outdoors.* If

*L. W. Neubauer and R. D. Cramer, "Solar Radiation Control for Small Exposed Houses," *Transactions ASAE,* vol. 9, no. 2, 1966.

deciduous trees are used for shading, most of the warmth of the winter sun can also be saved for heating. Placement of such trees is, however, crucial, since even deciduous trees intercept thirty to fifty percent of the winter sun's heating potential. Whenever possible, city planning policies should encourage the planting of large

Narrow streets

deciduous trees on the east and west of buildings where they will provide useful shade without shading south windows in the winter.

In neighborhoods with excessively wide streets, the beneficial effects of tree planting are considerably reduced because street trees cannot adequately shade them. For example, it has been shown that in this climate, urban areas with well-shaded streets often have maximum summer highs five to ten degrees lower than poorly shaded areas. For this reason narrow streets should be encouraged. At present, Davis streets are often too wide. Their design emphasizes parking, traffic flow, and minimizing the response time of emergency vehicles. While all of these are important concerns, they have to be weighed against the advantages of moderating microclimates and reducing energy needs. Because of the high cost of paving, 50 cents to $1 per square foot, builders generally favor narrower streets which reduce paved area and allow home builders to create more and larger lots. In an analysis, a typical Davis subdivision showed at $900 savings on a $7,000 lot when streets were redesigned to reduce their total paved area. Narrower streets also benefit the city's tax base by increasing taxable land and reducing maintenance costs.

Wide street

A NATIONAL GOAL: FIFTY PERCENT PER CAPITA REDUCTION IN ENERGY USE

If simple energy conservation techniques, such as those developed in writing the Davis energy conservation code, were widely applied to both new and existing structures, a vast watershed of electrical generating capacity and natural gas reserves could be released for other uses. But between this insight and its application lies a gulf of ignorance. The average homeowner, building operator, and even most contractors, architects and engineers know little about the thermal performance of their homes. The present problem of energy conservation in buildings is much like the problem of soil conservation once faced by agriculture. Farmers, it was found, knew very little about simple techniques of soil conservation. Through the educational programs of various agricultural extension services and a variety of economic incentives offered by the Soil Conservation Service, farmers were able to reduce soil erosion, prevent flooding, increase yield, and in general use better and more wisely their land's resources. An "Energy Conservation Service" could be organized along similar lines to advise builders and homeowners of the special needs and opportunities of their microclimate and to suggest proven ways of reducing energy waste. Cost-sharing, and, most important, on-site consulting could all easily fall within the mission of such a service.

Our study of Davis houses indicates that existing houses could meet or exceed the new code's minimum requirement with an investment of between $500 and $2,000. Each dollar spent to upgrade existing houses would save twice the energy produced by investing the dollar in constructing a new nuclear generating plant, such as the proposed $1.1 billion Rancho Seco II facility: this comparison is based on first costs only. It becomes even more compelling when one realizes that an increment of energy saved is saved forever, whereas increased generating capacities require constant supplies of fuel and maintenance, items whose costs can only increase.

These figures indicate the enormous potential for energy conservation in new and existing buildings. I am convinced that a fifty percent per capita reduction in energy demand for heating and cooling would be a reasonable ten-year goal for an adequately funded Energy Conservation Service. Such a program would have profound economic implications. It would help to stimulate the severely depressed construction industry. A properly conceived program could put thousands of unemployed people to work. There are many small businesses and underemployed small contractors engaged in the housing construction and remodeling business who have the necessary skills to perform and work and benefit from the program.

The nation should set the goal of reducing its overall energy use by fifty percent in the next ten years. With proper distribution of existing capital and know-how, an energy conservation program for buildings could easily achieve this reduction in energy use for heating and cooling. Similar changes could be made in other areas not discussed above if proper legislation were adopted to facilitate the change. For example, it is generally recognized that commercial buildings are excessively lighted; general maximum lighting standards could be set to curb this wasteful situation. In addition, limits could be set on decorative and advertising lighting without affecting our well-being. A heavy sales tax based on weight for new large automobiles would encourage smaller, more efficient automobiles. Improved city planning and street tree planting programs in our cities could improve the microclimate and thus create large additional savings in heating and cooling. Better bicycle facilities would encourage the use of this most energy-efficient means of transportation.

A study of appliance use we performed in Davis revealed that a large reduction in energy consumed by appliances could be achieved through the use of more efficient appliances and consumer strategies. In this area, legislation requiring the efficiency ratings on appliances would aid the consumer in choosing the most efficient units. The same tactic could be applied to houses; the landlord or salesperson could be required to make available to the potential renter or buyer the previous two years' utility bills. This would increase the marketability of energy-conserving structures. Building codes designed for local microclimates could greatly reduce the energy demand of new buildings. Special incentives should be offered to encourage industries that use large amounts of hot water or heated air to utilize proven solar energy technologies for preheating hot water or for drying material. The list could go on and on.

If our society is to compete as a manufacturer and trader of products on the world market, it must improve the energy efficiency of its entire system. The only realistic way to resolve our short-range energy problem is through energy conservation. There is no need for spectacular breakthroughs in this field. Many small and seemingly insignificant improvements in the energy efficiency of our life-support system will add up to some big savings. Yet there is one major feat that must be accomplished. That is: we must clearly recognize our situation; we must set clear, strong goals for energy conservation; and we must be willing to commit a large portion of our creativity and capital to the task. It is cheaper to save energy than to generate new energy.

The Systemic Consequences of Centralizing Energy Production Systems on Modern Societies

Kenneth E. F. Watt

Remarkably, given the great and growing interest in systems analysis in the modern world, it is only within the last two years that a systems point of view has begun to be applied to the impact of energy systems on society. This is particularly remarkable because with only a small amount of reflection, it becomes obvious that the impacts of energy use on modern society are so pervasive that it is only reasonable to expect a wide variety of systemic effects.

It will be easier to grasp the significance of various components of the argument if a flow-chart of the major postulated effects is explained as a background first. Figure 1 represents a hypothesis about the way in which the degree of centralization of energy production systems affects other aspects of modern societies. Two powerful causal pathways operating to foster centralization of energy production are the economies of scale made possible if very large quantities of energy can be delivered to large customers (1), and the availability of large quantities of energy at localized sites, making for low costs per unit (2).* Centralization of energy production creates five causal pathways. In the first of these, low prices to large customers encourage them to use more energy (3). This, in turn, increases the national fossil fuel use rate (4). Since the United States is now running out of cheap domestic fossil fuels, this in turn increases the volume of exports of other raw materials to pay for imported OPEC crude oil (5). These exports include wheat, corn,

*A causal pathway is a flow or path of causes from initial cause to final effect. Figures in parentheses in the text refer to the numbered pathways in figure 1.

FIGURE 1: THE SYSTEMS CONSEQUENCES OF CENTRALIZING ENERGY PRODUCTION SYSTEMS

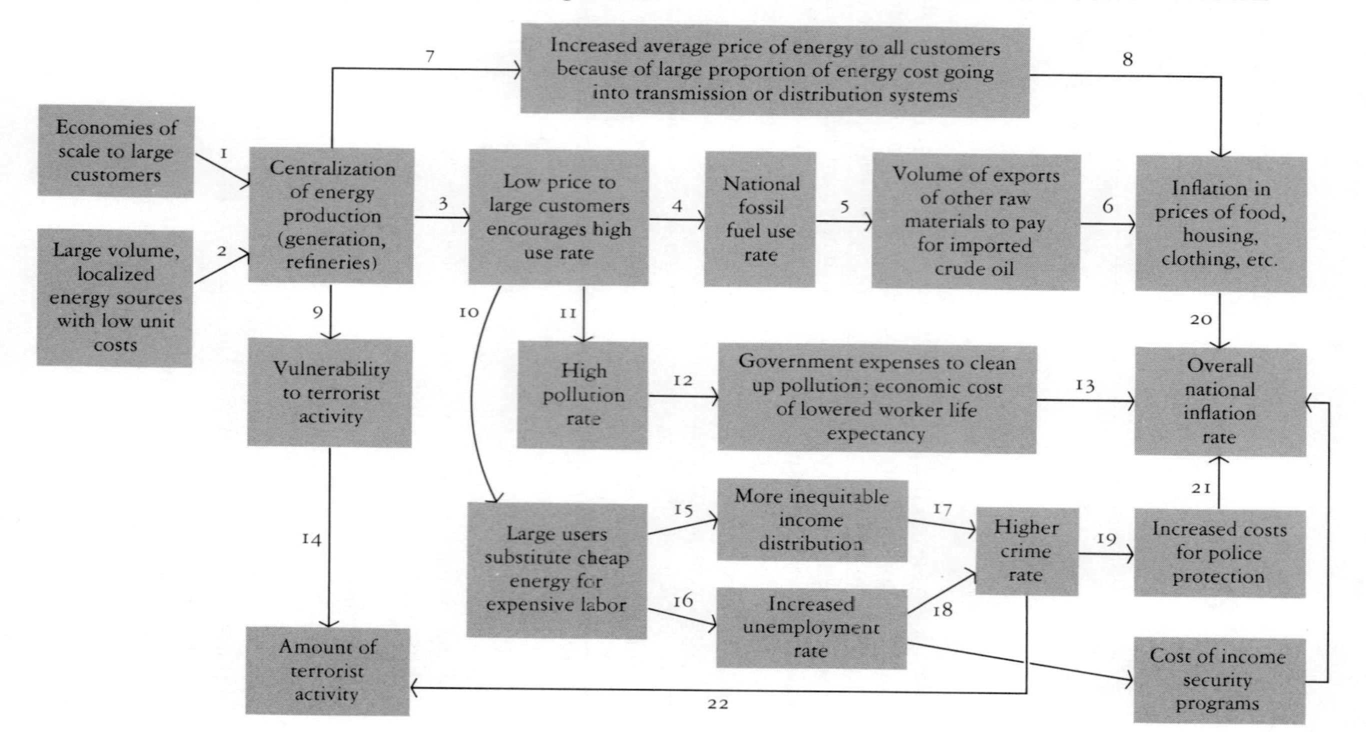

soybeans, all wood products, textiles, and ores and metal scrap. Running down domestic reserves leads to price inflation in those commodities, and everything made from them, including beef, housing, and clothes (6). In the second causal pathway, there is inflation in the price of energy from central production sites to all customers, on average, because of the great overhead in distribution and transmission costs (7). These costs also increase inflation (8). The third pathway generated by centralization, and low costs to large customers, is a high pollution rate (11), which leads to both government expenses to clean up the cost of pollution, and huge dollar costs due to premature death in the populations exposed to pollution (12). This also contributes to the overall inflation rate (13). The fourth effect of the low energy prices following from centralization is the substitution of this cheap energy for expensive labor (10). The high profits made possible allow some people to become very rich, thus increasing inequitable income distribution (15). Also, since some of the displaced labor will not be able to find other jobs, cheap energy promotes increased unemployment rates (16). These two effects lead to higher crime rates (17, 18). This, in turn, creates increased costs of police protection (19), which in turn also contributes to the overall inflation rate (21).

TABLE 1

COST AND USE OF ELECTRICITY IN RELATION TO SIZE OF CUSTOMERS

	Cost of Electricity ($/kwhr)		1974 Price / 1950 Price	1974 Use / 1950 Use
	1950	1974		
Residential or domestic customers (A)	.0286	.0280	.987	7.93
Commercial and industrial light & power customers:				
Small (B)	.0269	.0285	1.059	7.56
Large (C)	.00986	.0155	1.572	4.85
C/A	.34	.55	1.59	.61
C/B	.37	.54	1.48	.64

SOURCE: Data for calculations from table 897, *Statistical Abstracts of the United States*, 1975.

But the increased crime rate also produces increased terrorist activity (22). This is important, because the fifth causal pathway triggered by centralization is increased vulnerability to terrorism (14).

To this point, we only have an elaborate hypothesis, which would mean nothing unless each and every one of the twenty-two causal pathways could be shown to operate, using statistical analysis of available data. A simple table shows that energy prices are lower to large users, and also that changing the energy price changes the use. Table 1 shows that since 1950 electricity rates for large commercial and industrial customers have been between 34 and 55 percent of those for residential or domestic customers. But when the rates for large users were increased faster than for small customers, as they had been by 1974, this made the rate of growth in use much smaller for large customers than for small customers. This proves that the amount of energy used is very price-sensitive, a phenomenon that can be demonstrated by a wide variety of different sets of data or types of analysis. The relationship between inflation in food prices and United States energy prices is shown in table 2. The

TABLE 2

ENERGY PRICES IN RELATION TO FOOD PRICE INFLATION

	1971	1972	1973	1974
Cost of U.S. crude oil imports (billions of dollars)	1.88	2.61	4.69	16.48
Exports and shipments of wheat (billions of bushels)	.63	1.19	1.14	1.10
Old crop carryover of wheat (billions of bushels)	.86	.44	.25	.25
Weekly food cost for a couple with two children 12–18, as a percentage of spendable weekly earnings of a worker with three dependents	40	39	45	47

dramatic increase in wheat exports beginning in 1972, to pay for the sharp increase in crude oil imports which began that year, ran down domestic wheat stocks, which in turn produced a sharp increase in the proportion of the disposable income of families going into food. Comparable tables can be produced for clothing and housing.

Only in the last two years has information become available about causal pathway 7, the effect of centralization on energy costs. Amory Lovins concludes that for small electricity customers, only 29 percent of their electricity cost was for electricity; the rest was for transmission and distribution equipment and its maintenance.* This will be one of the economic driving forces behind backyard solar or wind-electricity systems.

Perhaps the most surprising pathways in figure 1 to most people will be those connecting energy to unemployment, crime, and police costs. However, these pathways are remarkably clear in statistical analyses. Of the developed countries, for years, the USA and Canada have been first and second in both energy use per capita and unemployment rates. France, Italy, Japan, and Norway, with sharply lower energy use per capita, have always had sharply lower unemployment rates.

To indicate the degree of the correlation between unemployment and energy use, a new set of unemployment figures released by the Organization for Economic Cooperation and Development in Paris was used. The organization recalculated the national figures produced by twelve of its member countries to make them comparable. Ten of the figures were published in the January 1, 1977, *London Economist* (page 72). These unemployment rates, for the third quarter of 1976, were treated as dependent variables in a regression analysis, with energy consumption rates per capita in 1973 being the independent variables (the time difference is reasonable because of the long lags in socio-economic systems). The analysis showed that 54 percent of the variance between countries with respect to unemployment rate was accounted for by energy consumption per capita. Since energy consumption per capita is a measure of the rate at which cheap energy is being substituted for more expensive labor, which in turn is affected by energy price, this suggests that unemployment rates are in fact related to energy prices. The 54 percent figure is particularly remarkable when we consider some of the other factors that would be expected to mask this relationship, such as the economic crisis now facing the United Kingdom and Italy. Both of these countries had much higher

*Amory B. Lovins, "Scale, Centralization, and Electrification in Energy Systems," paper presented at Oak Ridge Associated University Symposium, Future Strategies for Energy Development, October 20–21, 1976; to be published.

unemployment rates than those that would have been expected from the equation. Norway and Sweden had lower unemployment rates than would have been expected from the fitted equation, but this may be explained by the political systems in these two countries.

In short, many different types of data analysis and simulation could be brought forth to suggest that the flow chart in figure 1 does constitute a realistic description of the effects of centralization. Suppose the flow chart is totally realistic. What does this suggest about the prospective consequences of an alternate pattern of social organization?

Suppose instead of homes getting almost all their energy from utilities, they got most of it from rooftop or backyard solar or wind collectors, with storage systems for energy under the floor, or in an insulated room on the main floor. What would be the consequences? First, the price would be comparable, given the increase in the price of fuel that is certainly coming as fossil fuels become more scarce. Second, there would be less of an economies-of-scale effect, to encourage society collectively to use energy at a higher rate. On the other hand, needing more energy would lead to the cost of more wind and solar collectors, so that a homeostatic (negative feedback) effect would tend to promote energy conservation. The inflationary effects of the current situation, in which high energy use implies high raw materials exports, would be ameliorated. The more wind and solar energy were used, the less pollution there would be. The less premature death there would be, consequently. The more society can convert to solar and wind energy, the more sparingly energy will be used. This implies that high technology will focus on efficiency, not power. Further, there will be less substitution of cheap energy for expensive labor, and hence less unemployment. Cheap energy, which is an important stimulus to high inflation rates through three different causal pathways, will be eliminated, so the inflation rate will be lower, all other things being equal.

Finally, this entire situation is extraordinarily ironic. Centralization of energy production, and the resultant cheap energy, has been pursued by this society as an absolutely necessary prerequisite for a high rate of economic growth. Americans have become so culturally programmed to believe in this causal association that we couldn't even notice the interesting fact that all other countries had much more expensive energy than we had, particularly relative to wages, yet almost all had much higher rates of economic growth (amongst the developed countries). Finally, as Sweden and Switzerland have surpassed our GNP per capita, we still haven't grasped the significance. But in the future, rapid increases in the price of our fossil fuels, and enormously expensive nuclear energy, will drive us to decentralized solar and wind technologies. The result will be an improvement in our rate of economic growth. Probably even then, few people will be able to figure out why it will have occurred.

Nuclear Energy—
An Appropriate Technology?

Dan Whitney

Intermediate technology is a concept that is foreign to most people in this country; but then those of you who are here and involved in applying the concept already know that. As an example of where the concept stands today, I will relate to you a story of my first involvement with intermediate technology. Upon accepting the invitation to participate in this panel discussion, I thought it imperative that I read Mr. Schumacher's book *Small Is Beautiful.* Going to my local bookstore, I asked the clerk if they had the title, to which he replied, "Yes, it's right here in standard paperback, or if you would like, we also have it in a larger size."

I was amazed! The same book, both in paperback, with one version larger and having a fancier four-color cover. I selected the smaller, cheaper volume and mused about the viability of a society that can absorb the overhead of producing two like books, distribute them, and do it so everyone makes a profit.

"Energy." A single world that is many things to many people. Since we all use energy, many have determined that they are experts on the subject. We use energy, whether we know it or not, want to or not, every second of our lives. It permeates our American way of life. Is it good? Do we use too much? Do we need more? Can we use it more efficiently? Where is it going to come from? Tomorrow? A century from now? And my great-grandchildren, what will it mean to them?

The questions far outweigh the easy solutions, or even many of the known solutions. I, as a representative of the energy industry, cannot tell you that a solution to all our problems even exists.

Rather, I can confirm what you already know, that we live in a finite world, that growth has been ingrained in us, and that our need for "growth" is continuing at a time when we know that crisis looms. The time is coming when demand and consumption will both exceed the available resources and supplies. At that point one can envision a disaster of unbelievable proportions.

What, you ask, is the role of the energy utility in this equation? Certainly it is not to go along unheeding the signs about us. Yet, our charter is to satisfy the needs and demands of our customers. It is *not* to be the instrument of social change. I can agree with you that social change, or better, re-education, is the only way for the United States to achieve self-reliance and eventually freedom from external concerns. Education is the avenue to effect any meaningful change in energy policy and practices. But now let us look at where we really are.

The development of energy systems within this country has had to follow two sets of laws. The first being the laws of physics and thermodynamics, the second, those imposed by society.

The laws of thermodynamics cannot be changed simply because one does not like the influence they have or the processes they define. Yet, by the application and reapplication of what we have come to call technology, the forces of thermodynamics have been put to work for us. For example, the steady evolution of ever higher levels of technology has given us modern steam electric-generating stations with thermal efficiencies exceeding forty percent, compared to about ten percent in the 1890's. But a similar improvement in the next eighty years should not be expected, no matter how much we might desire it. The second law of thermodynamics simply does not promise similar increases. These laws, and the realities of economics, combine to hold even high technology to a limit. I mention this as an example to suggest that a reduction in the level of technology cannot be imposed without a cost, and depending on the magnitude of the application, the cost may be so great that it must be considered a luxury to accept the resulting inefficiency.

Decentralization of energy is such a luxury. The costs would be enormous by any measure. Today's energy systems were established at a time when economics was the primary criterion. If it could be done at a reasonable cost, and provide a useful result, it was done. Unfortunately, there are examples of situations where this was carried to an extreme, where utilities were high-handed or aloof, but the fact remains that today we enjoy a diverse and strong energy supply.

To discuss the disadvantages of decentralization, regional and dispersed energy systems, let us look at several examples of how advantageously the current centralized system serves the country. This discussion addresses the situation as it is. I am not recasting it, our society, or philosophy into some form which might be amenable to some proposed solution.

The West Coast of the United States is tied into one large energy grid. There are several large gas, oil, and electrical lines tying the geographically diverse areas together. As we all know, in 1976 California received only about one-third of its normal rainfall. Normally a large portion of northern California's electric energy is produced locally by hydroelectric generators. Last year, this power did not exist. Thanks to the intertie to the Pacific Northwest, we have been able to take advantage of the unusually wet year they had. Water that would have simply gone over the spillways was instead run through turbines, and the excess power thus generated transmitted to California. This meant we did not have to burn millions of barrels of oil within our valleys and basins, or as an alternative, reduce the level of economic activity.

This winter has also been unusually dry. Not only for California, but for the entire West Coast, including British Columbia. So that we may all save the meager reserves of water that do exist, fossil and nuclear electrical generation, excess to California requirements, has been dispatched to the Northwest. This aids them in maintaining adequate water to support their annual agricultural and energy requirements.

We are all aware of the extended and severe cold wave that enveloped the central and eastern portions of the nation early this year. The lack of natural gas developed to a crisis level. Business, education, and normal routine were reduced to a standstill. (It should not go unnoticed that natural gas, and its production, distribution, utilization, and cost, is the most regulated form of energy in this country.) Yet the impact of this cold wave would have been even worse had not distribution been shifted from the West to the East. The effect on the West Coast was hardly noticed, since during the period in question, conditions were mild, and adequate reserves of alternate energy were available.

The following is one final example of the benefits of diverse, intertied, and high-technology energy systems. A plan currently is under consideration to operate, for just the summer peak, a natural gas–fired thermal electric plant, located near Los Angeles, with gas allocated to northern California. This would help provide the energy which otherwise will be unavailable because of lack of water in northern California streams and reservoirs. The Los Angeles boiler would normally be unused, since it cannot burn oil in the L.A. basin and it does not have a natural gas allotment. These examples of problems solved by large and complex energy systems would be enormous obstacles to the very existence of the regions involved, were it not for the energy transmission networks.

Realizing the interrelationships between the American way of life and energy, we must now address certain individuals, and small groups, which would change the order and the relationships that exist in industrial America. With the advent of commercial nuclear power came a public licensing and approval process more involved and accessible than any before. Accessible in that it specifically makes provisions for the general public to be an active participant in the siting, construction, and operation of the power plants.

Nuclear power provides the opportunity, and the vehicle, to those who would impose a new set of priorities and values. At the risk of stereotyping, these persons are affluent, educated, and in positions which allow time for idealistic thought and the utopian adventure. By making symbolic commitment to their cause (examples which come to mind: "I reduced my electrical consumption by 25 percent last year" . . . "My new 4200-square-foot house will derive 80 percent of its heating from solar power" . . . "We have turned our thermostat down to 65° and wear sweaters around the house" . . . "I sold my old eight-cylinder Chevrolet and bought a new four-cylinder Porsche"), these people show their sincerity and opposition to that evil of American society, growth. Their ready solution to the "energy problem" is simply "conservation" and "develop renewable alternative forms," for example, wood, solar, hydroelectric. They truly believe that the world is committed to a helter-skelter rush that will exhaust all of its God-given resources and beauty. How do they rectify this situation? They would simply have to stop the "evil" growth. I feel that since all men are created equal, only education and due democratic process shall be allowed to effect any requisite change. If growth is the evil, then we must accept that fact, and limit growth. If this means we have to eliminate our industrial energy system, then so be it. But only if this is what society really wants. Not building the necessary nuclear power plants is one very effective way to stop growth. This is exactly what many persons are trying to accomplish. My request of them, is do not attempt to stop nuclear power with fears and frightful accusations. For in this case, no matter how moral the cause or intention, the end does not justify the means! Today this small elitist minority is manipulating the tools given for public involvement to effect their singular view of how our society should work and live. They are not the ones who will be made to suffer the consequences of their actions. The consequences will be suffered by the disadvantaged and those who' must work for a living. For these elitists expect to be exempt in their newfound positions regulating this reordering of our society.

If there are any doubters as to what I have just related, they need only look into the motivation, purpose, and support behind the recently proposed "Nuclear Safeguards Act." Fortunately, Proposition 15 was soundly defeated by the people of this state.

The chapter in *Small Is Beautiful,* "Nuclear Energy—Salvation or Damnation," is an excellent example of the inaccurate and biased critiques done against nuclear power. Typically these amount to a few words or sentences, lifted out of context, strung together with "concern," "faults," and ignorance to "prove" the author's accusations or thesis. No amount of fear, accusation, or ignorance will change the fact that nuclear power is the most environmentally acceptable alternative to the energy crisis confronting this nation today. (Let's be realistic, have you ever tried to develop an environmentally acceptable method for covering the earth with solar collectors? Twelve square miles of collectors would be required to replace a single nuclear power plant.)

But enough of this. The question is, "Is there a solution to this energy dilemma?" The answer is yes, there is a solution, but as usual, not everyone is going to like it. The solution I recommend is going to take time, and will require two different programs: (1) utilities must be deregulated and rechartered to serve only their customers' needs; that is, they are not to encourage growth for the sake of profit; and (2) re-educate the users, the customers, to reduce their demands and needs, to improve their internal energy efficiency, and to make use of viable alternative and passive sources of energy.

Government regulation is a major problem. It seems that anyone who has ever operated a light switch or bought a tank of gas feels he is an expert. In contrast, if you have any real practical experience you are labeled as prejudiced, biased, or simply one having a "conflict of interest." You are therefore ineligible to participate in the regulation process. One need only look at the proliferation of committees, agencies, and commissions in state and federal government to find numerous blatant examples. In general, none of these groups contribute anything to the betterment of mankind. They are overhead, a luxury to be exact, that only a very affluent society could even tolerate. The risk we run is that their incompetence and delaying actions will place this country in jeopardy of losing its very lifeblood, its energy.

This is not to say the utility industry is without its problems. Utilities are people, people are human, thus the order of society will always be necessary. The laws no utility can afford to break are those of economics. Economics is the great equalizer and regulator of this industry.

Re-education of energy users is a demanding and critical concept to the success of this program. Left to any single group it would likely become polarized, ineffective, or self-serving. Since we are seeking only truth, we must, therefore, turn to thermodynamics, economics, and the principles of application, such as environmental acceptability. These truths lead to the obvious and best solutions for a finite world. To allow education for the sole purpose of redirecting society reeks of a dictatorship. Our society has been based on the fundamental that truth would create an environment in which everyone would have an equal opportunity. This principle has served us well for over two hundred years and we have no reason to abandon it now.

What then of intermediate technology? It certainly has a place in developing nations, and as Schumacher says, is much preferred to imposing an expensive and high-technology solution on the unwitting. But it is important to look back into our own past and realize that we made use of intermediate technology during our nation's formative years. Likewise, we should expect our emerging friends to use intermediate technology to develop a life style similarly complex. I do not believe it is human nature to be satisfied with "intermediate success." There will always be someone with more wants, needs, desires, and ability. Someone who can afford to seek the rewards a higher technology can bring him. Do you need an example? Then look at the United States and Europe of today. Who really believes someone else will learn from our mistakes?

Where does all of this end? At the present we are extrapolating our consumption to infinity, and we do live in a finite world. Pessimistically, I have the feeling that the rich will get richer, the poor poorer, until only a few elite rich remain. At that time this earth will come into equilibrium, demand will equal the supply of renewable resources, and the need for growth for growth's sake will have been purged. Man will again be at peace with this earth. What kind of a man will this be? Will he be the peaceful soul we all say we want to be? Or will he be a ruthless and sly individual dependent on himself and his power to sustain his survival? I suggest the latter. What role does intermediate technology play in all of this? Intermediate technology is not "the" solution, although it may well be part of the general solution to the problems of mankind. There is no single panacea to the ills we have wrought upon ourselves.

Imposing programs like intermediate technology involves more regulation, control, and government. All of which are terribly inefficient and dehumanizing. It is far better that we return to the free market value system. Any proposed activity must be evaluated as to its benefit over its entire life cycle, with appropriate credit being taken for its true costs. There are no easy solutions to the problems we have produced. Turning our backs and looking for an easy way out will only increase the magnitude and hasten the calamity.

Discussion

AUDIENCE PARTICIPANT: The main emphasis of this conference has been that it's not so much the goverment institutions that have to be changed, but the values of the people. I have confidence in the people here although I'm sure a lot of us still have cars that we tend to use more than we should. I don't, however, have confidence in most of the people in society and I feel that the government must take a role and do some very politically unattractive things including imposition of a high tax on each gallon of gasoline consumed. This would appear to be inequitable, that's true; this could be compensated for by reimbursing the money according to income as stated on your income statement filed each year. This would have the effect of reducing gasoline consumption and rewarding those who don't use gasoline since you're not paid back according to how much you use, but according to your income. As far as I can see this proposal would be highly unattractive to the oil companies, but anything to reduce oil consumption would.

EMILIO VARANINI: I think that that type of proposal is totally academic at this time. And the reason I say that is I think a lot of people are tired of various proposed solutions which fall on the end-use consumer. We have a series of propositions in California where we are trying to put in insulation standards and energy-conserving standards in homes; we are trying to put in energy standards on appliances and so forth and so on. The public is tired of bearing the cost of these types of standards and what they are telling us now is to select off larger technical options rather than trying to coerce certain kinds of activities by the end users. It has been proposed in geothermal leasing, it's been proposed in several other activities, and it's just politically unacceptable.

I think that you're going to find that the pricing structure right now is on a fairly strong climb, particularly when the federal government deregulates domestic oil.

AUDIENCE PARTICIPANT: Well oil is artificially high and the people who profit from that are a very select few. At least this way the money would go back to the people. It's taxed and then it's given back according to income.

EMILIO VARANINI: I understand the principle but I just don't think it's politically feasible at all at this time. If you have a democratic form of government you have certain political constraints.

KENNETH WATT: I think the last comment and some of the chuckles, retorts, and remarks I heard around the room this afternoon are the essence of what the problem is all about. There is a deep-seated, widespread feeling in this country that what more expensive energy means is that the rich get richer, the poor get poorer, and the poor get screwed. And that's our problem. That belief. If we can get rid of that belief we'll solve our problems. For those of you who believe that high energy prices mean that the rich get richer and the poor get poorer, you're going to have to explain to me that when we have higher energy prices obtained in all other countries, that we don't get the situation you're all talking about. In lots of countries where energy is sold retail for a lot more than it is here relative to wages, income distribution is more, not less equitable. Let me put it to you the following way. Suppose you are a small company and you are trying to gain inroads into the competitive position with respect to big oil companies in providing energy for the public. What is going to make it easier for you to get going? High energy prices relative to wages or low energy prices? Well I know, I talked to the little solar energy companies, and we have our economic thinking all fouled up. We somehow think that life gets worse for everybody if prices go up relative to wages. And it's not true. First of all, we've got wages all mixed up with buying power. It's buying power that determines how your life is going. And you haven't thought through the fact that yet wages are high and energy prices are low, that you're going to have a constant substitution of cheap energy, that is substitution of automation and mechanization for expensive labor. And guess who it is that's going to be replaced the most. The poor people with no skills, so it's low energy prices relative to high wages that is the enemy of the poor and not the other way around. And until we in the United States can get that figured out, we're just going on into hell, let me tell you.

AUDIENCE PARTICIPANT: I have a couple of questions about nuclear power. I have never heard the utilities industry address these directly and so I would like a couple of answers. The first question and part statement is the process of uranium enrichment consumption involves the consumption of large quantities of electricity and so the conventional nuclear reactors that just use enriched uranium are essentially a process of transporting energy in nuclear form from places like Oak Ridge where it's enriched with hydroelectric power to other areas of the country. So the only really efficient form is that breeder reactor that produces plutonium. And I was just wondering how the utility companies could morally justify using a fuel one microgram of which is enough to cause lung cancer. The other question is, what confidence do the utilities have that there will continue to be agencies capable of safeguarding the nuclear wastes during the period of their half-lives, some of which are a quarter of a million years?

DAN WHITNEY: About the amount of power that's used for these enrichment light water reactors, in the case of Rancho Seco Nuclear Power Station, we used about 3.4 percent of the output of the plant in the enriching process, actually less energy that what it would have taken to transport coal the required distance for an equivalent-sized power plant. We can show you the receipts on that if you are interested. The interesting thing is many people feel like, is that all you're doing is transporting energy produced at Oak Ridge to here and that these plants consume more energy than they produce? In fact, that's not true. That is true if you go to the navy nuclear submarines which have a very small reactor running at high concentrations of enrichment and in those cases they do put more energy into them than you can get out. But of the enrichments we use that's not true. Now as far as the morality of producing plutonium by the utilities, that is certainly a question that is before our society today. But whether or not you address that question, it's strictly in the content that the utility does not eliminate the problem of plutonium. Plutonium has been around for a long time; it's around in enormous quantities; there is I think some six thousand tons of the material that has been dispersed into the atmosphere as a result of weapons testing. I'm not giving you these numbers with the idea of telling you that therefore it is safe and we should continue to do that; certainly not. But I am attempting to tell you that the figure you quoted as a tenth of a microgram being enough to cause a cancer is a figure that is very much in dispute; in fact that has recently been disproved with respect to particle theory. But the point is very academic if we want to argue how much plutonium it takes to cause a cancer. You have to first of all get that plutonium into the environment before it's going to be able to do that and we simply do not put the plutonium into the environment. The very small amounts that might leak out of a waste storage facility are extremely small. Plutonium is not the kind of material that likes to reside in the atmosphere; it is not really available to get into the lungs.

Your other question dealt with the storage of waste with half-lives as long as a quarter of a million years. Quite frankly, that simply is not true. The longest half-lives of the fission products in the reactor are thirty years. Ten half-lives gives you less than a tenth of one percent of the material. Now the material you're talking about is plutonium and there are trace amounts of plutonium left in the reprocessed waste. And as a result of that with its 24,000-year half-life, ten half-lives gives you a quarter of a million years. Most people like to conservatively apply a factor of 20 instead of 10 and so we get a half-million years that the stuff is "deadly radioactive." Well, it's simply not true. In fact, if you take nuclear ore and refine the ore, put it into a reactor and operate that reactor, reprocess it, and store the waste, you will find that in a period of less than 1,000 years, from the time that ore was dug out of the ground, the waste material has less radioactivity in it than what it would have had if it stayed as ore and never been mined in the first place. So in a way we're really doing you a favor. We are removing radioactivity from the environment.

EMILIO VARANINI: I have a comment about this. Many know of Alvin Weinbert and his famous quote about the Faustian bargain.

His view of what's going to have to happen for us to have any real sensibility with plutonium in the general stream of commerce is that we're going to have to have ten centers in the United States of about forty square miles each with two breeder reactors and a multiple number of burner reactors, and that those sites will have to be hardened and guarded and operated by an armed force with what he equates to the Strategic Air Command. And the reason for that, and he's an extremely pro-nuclear individual, it's his view that otherwise society cannot accept the security measures that would be necessary in terms of the potential of plutonium availing itself within the general stream of commerce. So I think that the one issue we really are facing in terms of letting genies out of bottles and so forth is whether nuclear is an intermediate-term technology or whether it's going to be brought up to a full-scale, longer-range technology. And that is still being debated.

AUDIENCE PARTICIPANT: My question is to Mr. Whitney and it's in relation to nuclear safeguards. In California they're building a supposedly safe reactor on an earthquake fault line and that worries me a bit. They assure us that it's going to have earthquake standards incorporated in it but I'm really unclear as to what those are. The other factor is in relation to the cooler systems on these reactors and the fact that the cooling systems are predominantly based on water and the fact that water seems to be becoming intermittently available in our state.

DAN WHITNEY: At the Rancho Seco site we have cooling towers and we only require make-up water for the evaporation in them. Now that's a relatively large amount of water in the sense that it amounts to about ten thousand gallons a minute which over a year's operation will amount to something on the order of sixteen thousand acre feet of water which in a normal year would be the amount of water that would fall on about ten thousand acres of valley land. So it's not an enormous amount that you're depriving someone else from. Now the plants you build farther along the coastline can make use of sea water and would make no major effects in the consumption or diversion of water from some other use. Now your question about two plants being built along the coastline near the earthquake fault, which is the Diablo Canyon facility. Those two plants are nearly fully completed and they have now identified a more capable earthquake fault closer to them than that for which they were designed. The design for that plant was the highest design in the country for earthquake standards when it was originally designed, which was far before 1971. In fact, when I went to work on the Rancho Seco project in 1970, they had already delivered the major reactor components to the site down there. They were far ahead of us in construction in Diablo Canyon. But, the earthquake design that is applied to a nuclear power plant is one of constructing the reactor and interconnecting piping and steam heat removal system. You construct that and then you test it or design it to tolerate an earthquake of a given magnitude. And the way that they do that is to put braces around the system to absorb the shocks of the various size earthquakes you design it for. In Rancho Seco we're designed for a quarter of a g horizontal acceleration. Now you want to separate the g number away from the Richter number because the Richter number has no meaning at a particular site. It only stresses the amount of energy released at the point of the earthquake so it does not address how much of that energy is there available as ground accelerations at the reactors. What they would need to do then is just increase the number of braces or supports to tolerate whatever number you want to put on it, like if you want to go to three-quarters of a g or one g, it can certainly be done and it's well within the economic realities of what the power is worth to do that.

AUDIENCE PARTICIPANT: My name is Richard Crayson, a county supervisor for the County of San Luis Obispo in which the Diablo Canyon plant is located. I wish to address a point that is perhaps not normally made but I think ought to be because I don't think we have to be talking about foreign countries and dependence and appropriate technology. We can look closer to home. The Diablo Canyon plant is nearly finished in our agricultural county, a rural county with 130,000 people, where our main economic base is agriculture, tourism, and government services equally distributed. We have very little industry, less than five percent, except for the Diablo Canyon plant and the Union Oil coke plant. At the present time most people in our county believe that tourism and agriculture and government services are our main economic base. But over the years, almost ten years now since Diablo Canyon has come in, and I'll give you 1975 figures, 22 percent of our county tax assessment comes from Pacific Gas & Electric Co., 49 percent of the local school district tax money is generated through the assessed values of the PG&E plant in our county. Now you talk about addiction, we are addicted to power in our county. It has come upon us in a way with those members of the board in prior years who wanted that power because they thought nuclear energy was good and did all the things that we heard it was supposed to do. Now we find ourselves dependent for about 25 percent of our income on that plant's existence, and 49 percent of the local school district which takes in almost half the county dependent on the taxes generated by that plant. I happen to be an elitist who worries about that kind of addiction because I see it as a very critical portion of the entire power structure, that a large corporation like that can come in by invitation ten years ago. But here we are, that plant is there and we are dependent on it now. We are just as dependent as Seattle was on Boeing, and Detroit on automobiles. I don't like that situation.

VIII

SUMMARY AND CONCLUSIONS

Gary Goodpaster

Summary and Conclusions

Gary Goodpaster

I was supposed to summarize this conference. What I'd like to do instead is simply focus on something not adequately covered in the conference, and that is the civil liberties aspect of "small is beautiful" notions and ideas. There is another reason for doing that—I'm here as a representative of the ACLU, and I think their viewpoint should be expressed. So I'd like to do that, and perhaps say a few things by the way of summary, and then simply open discussion for reactions, reflections, and summary from the audience.

Dr. Schumacher's speech, his major address of the conference, talked about civil liberties. He said something to the effect that we have to learn to live with necessities, and that was his basic message. For me it wasn't very concrete. But I think there is a concreteness in his message. This basic approach related to the limits that we face and what's going to happen to us; what sacrifices will come, what costs we'll bear—because now and continually in the future, we're going to face more and more limits; less fuel, less water, less food, and so on. The poor were mentioned, urbanization and mass unemployment, yet no connection was made to the consequences of their possibilities for civil liberties. I'd like to draw some of those consequences and I'd like to do it in a very personal way from my own experience.

All of my adult career as a practicing lawyer has been spent in trying to provide legal services to the poor. I started out as a kind of warrior in President Johnson's War on Poverty. I followed a kind of ideology which was: there was an enemy, the enemy was the government; big bureaucracies, most business; and racial and class prejudice. And I was going to use those lawyer's tools available to me to fight these things. I was going to use the very same tools that the people who ran the country used, the law and lawsuits, but I was going to use them to assist the poor. But the overriding ideology was that of community action, which says: the reason people are poor in this country is because they don't have power, and the only way for them to have power is for them to become organized. So my developed ideology was that I must use the law to organize the poor. I tried to devise lawsuits and organize community groups around those lawsuits. (It was very heady times for me. It is a thrill to think of yourself as acting in the public good and that you are a kind of white knight chasing evil villains.)

The first years of the legal services program for the poor resulted in many seeming victories. We lawyers deemed them to be really brilliant. But pretty soon there was a counterattack by the forces that had real power and most of those earlier legal victories were undone in treaties signed by the legislatures and courts.

Nevertheless, I learned some things from this war on poverty (which is now reduced to less than a skirmish) about the nature of poverty. I learned, for example, that there aren't many gains that are established when someone who is actually a part of the elite tries to assist people and do something for them that is perhaps better done for themselves. What I thought were tools to assist these people in fact may have injured them more than having left them where they were. To give an example: in Colorado one of the really bad problems was migrant labor housing. Colorado had a good migrant housing code. As a lawyer such a code gives you a ground to attack. You can say to a farmer about his migrant housing: Your housing is substandard; you have to upgrade it. If you don't we're going to shut it down. You are therefore going to be deprived of your migrant labor force, and you're not going to be able to harvest your crops. That is simple logic. In actuality, using that law and going after the migrant housing, the new result was that the farmers, instead of repairing the housing, simply closed it. Consequently, our clients, the migrant workers, had no place to live. They continued to work but their housing situation was much worse than it was originally.

Beyond this, I learned that some things which were seeming victories really made things much more complicated. One of the big victories for legal services lawyers was the establishment of certain rights for people who are going to have their property taken away by someone else, a creditor for example. One of the rights that has been fairly well established is that if you are going to have property taken away from you, you've a right to a due process hearing. Before they take it away they have to give you a notice, especially if the state is going to take it away, and in some cases, you may be entitled to a lawyer. It looks like a fine protection for the poor. But look how this ultimately operates in the credit market. The net result is that lenders, realizing that they will have to go to a hearing before they take their security back, increase the price of credit. They increase the cost of credit by increasing the rate of return that they get on it; consequently they price more poor people out of the market.

Aside from this serious consequence, when you have hearings, and when you have lawyers, you quickly have an arena where you can make the law more complicated. It is in a lawyer's interest to make the law more complicated. The lawyers do that to help their clients. But the net result for the poor was that they lost the credit market which they badly needed, and at the same time the law became more complicated. And in making it more complicated, developing a greater sophistication in the law, we lawyers were moving in what I would call an ever increasing, ever tightening gyre, an upward spiral of sophistication that was going up and up. In effect, by trying to bring fairness to people, by trying to bring fairness into the law as it affected the poor, we were simply stretching out the class structure, building the structure of expertise, making more and more experts. And when you make more and more experts with a vested interest in more refined knowledge, it does separate people and many begin to drop out.

Now you can see clearly this consequence, which I connect very much with the Small Is Beautiful Conference and some of the basic ideas presented here. We are in serious trouble because of the sophisticated nature of a lot of our knowledge. We are now becoming deeply dependent on experts and cannot do things for ourselves; and the more we use experts to solve problems, for many people matters are made worse. The way we use law and the way we use technology to solve the problems that we have may end in making things more difficult for people. This very effort to seek rights, to create rights for people, may in fact cost them more than it gets them.

I left that kind of law practice and now I'm a criminal lawyer. And I represent, for the most part, people I don't think you like. They are murderers, rapists, and robbers. Some are innocent, but most of them aren't. I represent convicted criminals on appeal. I see very clearly from my point of view that a lot of them are there because of the nature of our society. Because of mass urbanization, they are living in crowded conditions. Because of mass unemployment, they don't have work. Because of the character of our economy we stimulate desire for things that these people, for the most part, cannot acquire with their ordinary abilities and opportunities. We teach them to want things they cannot get legitimately. We've created a very sophisticated society without the need for full employment, because of automated machinery, and we require a social and educational discipline of people, many years of training and study to get into the system. Many people simply cannot meet these requirements and therefore cannot succeed in the system. Consequently, a lot of them simply drop out, and there's a major crime problem in this country.

I don't know whether or not there is an increase in crime in this country today, but I do know that there is a major increase in the concern about crime. The public is frightened, deeply frightened. I think the sign of the day of the future may be the locked door. Watch this next gubernatorial campaign (1978). It is now heating up. And you find, for example, that Chief Davis is saying nothing but law and order. Governor Brown is responding to the law-and-order cry, and he is sponsoring legislation which is strictly law-and-order legislation. Now I'm not critical of that, but I am trying to make a point. As a criminal lawyer I deal only with symptoms. I don't deal with causes. And none of the legislation that's coming down is dealing with the cause either. It also deals with symptoms. The legislation that you find is the kind that enhances police powers, increases electronic eavesdropping and surveillance, for example, and in other ways broadens police power generally. It is done because that is the only thing that is known to be done, and it satisfies an enormous political demand. "Give us security, make our streets safe, let's not have a murder a day in San Francisco, let us go out, let us not lock our doors, we're afraid." So we have this legislation, but the legislation won't treat the causes of crime. It only deals with symptoms. On the other hand, I'm really scared of the legislation because, from the point of view of the civil libertarian, every police power you grant is a taking away of some liberty you have. It increases the power of the state. We grant such powers because of the intractable problems and because we don't know what to do. We are enhancing powers that may ultimately cost liberties. The mass urbanization problem that we have and the mass unemployment problem that we have which are addressed by Dr. Schumacher's ideas, are clearly problems that in some way are going to lead to a diminishment of civil liberties for all of us.

Now I'd like to generalize from the civil libertarian viewpoint with respect to similar problems in foreign nations. If you look at the African countries and some of the South American countries and note the major cities which are created on a Western model and Western style, they seem almost to have come from the United States. Very modern, very attractive. They are filled with all these glittering discotheques and boutiques of desire that we generate in the United States. And they are attractive to people who move from rural areas to these towns. So people in the rural areas move to these towns and cities, and you find the cities fringed with shanty towns—people who are unemployed, people who have lost their roots and connections with their past and are now trying to be integrated into a new economy which does not yet exist, but are nevertheless attracted and hang on.

These populations are large and they may get larger. What I see coming—I think it has happened already—is that when these populations get sufficiently large and sufficiently restive, then their governments must take repressive measures against them, or, lacking that, some other force will utilize them to take power for itself in a revolutionary way. In other words, one has a volatile situation for an explosion and thus the possibility of military dictatorships in these countries.

We not only worry about what might happen to civil liberties for those people, it also has civil liberties repercussions for us. Because of the very nature of the process, because of the volatility and because of the problems that we have ideologically with the Soviet Union, these areas are disputed areas. All these poorer states are becoming client states. They're client states of ours and they're dependent states of ours. So we may have to take sides when a foreign explosion occurs and align ourselves with governments that may ultimately be quite repressive. You are all familiar with the events in Chile and the role our government played there. So I discern that mass urbanization, mass unemployment, the flight of people from rural areas, the loss of work and rewarding work in rural areas, lead directly to things which may call forth dictators, that may call forth much more repressive police measures, and many greater efforts to control populations. In our own country there may be a countervailing need to maintain a large armed force, to sell arms, and to do all those other things to maintain national security which are also going to be dangerous to our own civil liberties. I think it's simply going to happen that way.

Now so much for the side that I see in respect to the poor, or with respect to mass urbanization and mass unemployment. I want to touch on this notion of limits. We are tremendously interdependent, unbelievably so. Let me give you some examples. You know the switch, the little switch, the little rheostat in the power station that goes out and then everyone suffers. One well-placed gelignite charge on the aqueduct from Los Angeles to the Owens Valley and that city is going to be cut off from water. They have a ninety-day water supply in a reservoir and I suppose the aqueduct could be repaired. But it shows you how sensitive it is. A similar kind of charge placed at the base of the electrical transmission towers that run through Oregon, could cut off power for all of western Oregon. Because we are so interdependent, we are very subject to terrorists, and to terrorism generally. And this is, again, one of the major problems we are facing. The security checks that you go through in the airport as you know did not exist seven or eight years ago. It was a nice, free, easy walk into the airport. Now you've got to go through the magnetometer, and there may be other more severe security checks. They have a profile, and if you match the profile, whether or not you're a criminal, whether or not you're carrying a weapon or any kind of charge on you, they're going to pull you out of line, and they are going to search you. Examples of what can be done to us because we are so interdependent.

Dr. Schumacher's ideas addressed all these issues in a very different way. The most important statement of Schumacher is a moral statement which is: you have to change your way of life. You've got to change your way of living because your way of living and my way of living lead to these things that we are all so worried about. He is saying there must be a deep, moral commitment to some values that we all feel we've lost.

In trying to prepare, I asked myself why this mass appeal for Dr. Schumacher's argument. There were 1800 people here last night to hear him talk. He has a six-week tour in this country, and here's a man talking about technology. Why this mass appeal for technology? I don't think the appeal is for technology; I think that's a means. And the real solution is not technology, but it's the idea that there is a moral change that is wanted, a commitment to a new way of living. And the new commitment is the realization that civilization starts with us, it starts with each of us individually and what we do in our daily lives. We can no longer, if we ever really did, depend on a government, on a bureaucracy, or anyone else. In a way I think his philosophy says, "Save yourself." But I think there's something additional. It says that if you save yourself, there is a likelihood you'll save others. And the way to save yourself is to go back. Find some things that are genuinely satisfying to you. Find joy in work, do those things which gratify you. Most of your time is going to be spent working; it should not be spent unpleasantly; it should be spent happily doing what you want to do. It would be enormously satisfying if you could find that kind of work. If you are happy in your work, if others can be happy in their work, they perhaps will not have the same needs for the attractions and seductions of our culture.

This solution Schumacher offers is what he calls the appropriate technology or the intermediate technology. I've had some difficulties with the solution, but I think what he is saying is this: there is a way to live that gets back to natural living, to basic values, which is simple, which does not damage or destroy the land, which is satisfying and which is fulfilling, which is a way of life that requires all of us to work but to work in ways that will please us, to work in ways that perhaps the human being was designed to work. So he says, let us devise those instruments which we well have the knowledge to devise, which will please us to use, which will take care of the problems we are having and clearly identify them. This will take care of the problems of mass urbanization, of mass unemployment, and limits. It will take care of these problems by providing work, by not exceeding the limits or the carrying capacity of our environment, and by giving a kind of work that is satisfying.

Where do we go from here and where do we take off from? If Dr. Schumacher's basic idea is that a moral commitment is required first, then the place to begin is to examine one's own life, see what your way of living is, see what my way of living is, what its consequences are, where they go, what's going to happen if that continues. If our assessment results in a continuation of our present way of life and only more of the same, then that's wrong in his terms. It is a decision point. We can go forward, continue as we are, or change our ways of life. There is a necessity for a personal commitment here, and if you make this personal commitment to the way of life that he is talking about, everything else in some sense is going to follow. Whether or not the world is going to be saved is, in a sense, irrelevant because at least we will have lived a satisfying life, and that will be something. It will be very valuable indeed. And as he said, perhaps it is going to lead to something else. It will be one step at a time. So personal commitment is the very first thing.

A second step that Schumacher heartily endorses is that nothing is done without action. We talk and have two-day conferences—academics are really very good at that, and dividing and splitting hairs, and saying he didn't make this ethical point and that ethical point—but what is needed is to do something—to do something! To take action if you believe in it. And I think it is a moral belief that he's talking about. If you believe it and you commit youself to a new way of living then you must do something, you must live that way and then be the example to others of how to live that way, to show a satisfying life. Talk to others, share your ideas. Perhaps a local intermediate technology group could be started. Other groups could be formed along these lines.

Finally, it is very important, if you believe in what Dr. Schumacher says, that you must bring your belief into political form. It must be brought into the parties and it must be pushed as a part of the platforms. If we all believed in living the way Schumacher suggests, then our society would be very different. So we are talking about changing beliefs and changing them fundamentally. That's a very hard thing to do. The inertia of our institutions and the inertia of our beliefs, the momentum that they have, is unbelievable and fantastic. It may be possible to change them, but that will require that you live steadfast in your beliefs and seek to see them realized in your politics.

There's a story I really like that teaches something important about ideas, their limits, how much their power may reach, and I'm going to end with it. It's a Hindu story. It is said that there was a monk who was engaging in cosmological inquiries against the teachings of the master. In other words, this monk was questioning what the master said about the origin and nature of the universe. The monk wanted to know where the universe ended, where was its stopping point. And the master turned away several times very angry, beat him a few times, but the monk returned again and again and finally, the master, almost in despair, said, "Well, I can't tell you, I don't really know that. I'm going to refer you to a higher authority. I'll refer you to higher up, to the first god." The monk goes to the first god and gets a similar treatment, a similar response. There's no answer. Where does the universe end? Anger, rage, and then finally in frustration, "I can't tell you and I'll refer you further up." They have a bureaucracy and hierarchy of gods and then ultimately he is admitted to the divine, radiant, and omniscient presence of Brahma, the Chief and Head of all the gods. The monk walks into this fantastic and glorious assembly, and all around are radiant lights so blinding that he can't see. He walks before the throne of Brahma and very bravely, his hand up shielding himself from the bright lights, asks, "Dear Lord, Good Lord, I've asked my master, I've asked all the nether and lower gods where the universe ends, and they couldn't tell me, so now I ask you." Brahma, very quiet, very stern, finally gets up from his throne, stands, gestures to the monk and asks him to walk with him aside. They walk away from the assembled crowd so that they can't hear them. Brahma turns to the monk and says, "Well, I brought you away from all my attendants and gods when you asked me the question, where does the universe end, because I know you asked me because I am all powerful, all knowing, but I didn't want them to hear my answer. But, I don't know where the universe ends."

Now I don't know where these ideas end, but I think they are good, and if we did follow them, they would end well for us.